# Mining on a Small and Medium Scale

## A Global Perspective

*Edited by*
**AJOY K. GHOSE**

**INTERMEDIATE TECHNOLOGY PUBLICATIONS 1997**

Practical Action Publishing Ltd
25 Albert Street, Rugby,
Warwickshire, CV21 2SD, UK
www.practicalactionpublishing.com

© Intermediate Technology Publications 1997

First published 1997\Digitised 2013

The author has asserted their right under the Copyright, Designs and Patents Act 1988 to be identified as author of this work.

All rights reserved. No part of this publication may be reprinted or reproduced or utilized in any form or by any digital, electronic, mechanical, or other means, now known or hereafter invented, including photocopying and recording, or in any information storage or retrieval system, without the written permission of the publishers.

Product or corporate names may be trademarks or registered trademarks, and are used only for identification and explanation without intent to infringe.

A catalogue record for this book is available from the British Library & Library of Congress

ISBN 10: 1 85339 401 7
ISBN 978-1-85339-401-0 Paperback
ISBN 978-1-78044-548-9 Digital book

Citation: Ghose, A. (1997) *Mining on a Small and Medium Scale: A global perspective*, Rugby, UK: Practical Action Publishing https://doi.org/10.3362/9781780445489

Since 1974, Practical Action Publishing has published and disseminated books and information in support of international development work throughout the world. All print editions are produced and distributed via ethical and sustainable print on demand global facilities.

Practical Action Publishing is a trading name of Practical Action Publishing Ltd (Company Reg. No. 01159018 | VAT 880 9924 76). All profits are covenanted back to its parent group, Practical Action (Charity Reg. No. 247257).

The views and opinions in this publication are those of the author and do not represent those of Practical Action Publishing Ltd or its parent charity Practical Action. Reasonable efforts have been made to publish reliable data and information, but the author and publisher cannot assume responsibility for the validity of all materials or for the consequences of their use.

The manufacturer's authorised representative in the EU for product safety is Lightning Source France, 1 Av. Johannes Gutenberg, 78310 Maurepas, France. compliance@lightningsource.fr

# PREFACE

Small and medium scale mining occupy a special niche in the mining world and have a dominant share of the global mineral production. According to mining archaeologists, the art of small-scale mining evolved from the surface scratchings for red ochre in Bomvu Ridge in Swaziland around 40,000 B.C. to the underground mining for lead and zinc at Rajpura-Dariba in Rajasthan (dated to 1260 B.C.), to more organised mining around 300 B.C. reported in Kautilya's *Arthasashtra* and much later in medieval ages described in *De re Metallica* by Georgius Agricola (1550 A.D.). The perceptions on small, medium and large scale mining must have changed over the years. The transition to medium scale mining, as we perceive today, must have come about in the middle of the nineteenth century when energy from steam, compressed air and later electricity and also the chemical energy of explosives replaced to some extent the 'brawn' of the artisanal miner. Large scale mining entered the mining scene only in the twentieth century dictated largely by economies of scale to exploit ores of declining grade, and from greater depths. The mining industry presents today a variegated mix of small, medium and large scale mines. However, problems of definition abound as one attempts to carve out the specific domains of each sector. Be that as it may, small, medium and large scale mining, was, is and will always be important to mankind as the cornerstone of civilization, the very basis of sustainable society.

This volume collates the proceedings of the Global Conference on Small and Medium scale Mining (GCSM'96) at Calcutta which has been convened to direct the spotlight of attention on this vital segment of global mining industry. Interest in small scale mining has been heightened in recent years due to the increasing concern on its environmental impacts. We have travelled a long way, and so it would appear, from Jurica (1978); Taxco (1981); Helsinki (1983); New Delhi (1984); Ankara (1988): Calcutta (1991); Harare (1993) and Washington (1995) debating interminably on the positive and negative impacts of small-scale mining. The issues are certainly vexed and have merited consideration at such important fora as at the United Nation's Economic and Social Council (ECOSOC), the World Bank and the International Labour Organisation. This present volume is not intended to fuel the debate; it seeks merely to delve into the many dimensions of small and medium scale mining and to take stock of the emerging issues. A recent report from the UN Secretary General presented at ECOSOC has underscored once again the growing importance of the small scale mining sector world-wide with an estimated employment of over 6 million people. The sector contributes estimatedly to 12–15% of the global metal production, 35% of the non-metallic production, 10% of gold and diamond production and 20% of coal production. With a large female workforce and widespread employment of child labour, the small scale mining sector also poses very high health and safety risks. It has also rightly been identified as a major predator on the environment.

By and large, the contributions in this volume relate to *issues management* vis-a-vis small scale mining focussing on environmental, safety and health, social and gender issues. Medium scale mining *per se* has received scant attention, possibly because of the fact that there is a commonality of the issues amongst small and medium scale. There are excellent contributions sharing world-wide experiences and a few on technological options to assist the small scale sector. Collectively, the papers provide on a platter a multitude of issues and their solution relating to small scale mining which could hopefully form the basis of an agenda for action.

GCSM'96 and this volume could not be possible without the unstinted support and cooperation from many sources. The initiatives of the host, namely the National Institute of Small Mines, were ably reinforced by Small Mining International and Intermediate Technology Development Group. The Ministry of Mines, Government of India and the Federation of Indian Mineral Industries (FIMI) extended valuable cooperation. This publication has received funding support from International Labour Office; for this generous contribution we owe a debt of gratitude to Mr. Norman S. Jennings at I.L.O. My personal thanks are due to Mr. S.L. Chakraborty, Hony. Secretary General, National Institute of Small Mines and to his team for their inputs which enabled the publication keep its planned schedule and timing.

November, 1996  
Calcutta

AJOY K. GHOSE  
Convenor GCSM'96 and Editor

# CONTENTS

Preface   iii

## SECTION 1: SMALL AND MEDIUM SCALE MINING/ISSUES MANAGEMENT

Addressing Social Tensions in Mining: A Framework for Greater Community Consultation and Participation   3
*Mamadou Barry*

Small-Scale Mining: Time for Deeds Not Words   9
*Norman S. Jennings*

Strategies for Development of Small/Medium Scale Mines in Africa   17
*Pierre A. Traoré*

An Integrated Participatory Framework for the Exploitation of Mineral Resources in the Context of Sustainable Development — The Case of Namaqualand, South Africa   25
*Olle Östensson*

Policies for Artisanal and Small Scale Mining in the Developing World — A Review of the Last Thirty Years   35
*John Hollaway*

Gender Issues in Mining — Effects of Retrenchment   43
*Paramita Aich*

Legislative Framework in Indian Mineral Sector and Scope for Its Improvement vis-a-vis Small Scale Mining   53
*R.L. Bhatia* and *S.V. Ali*

Cluster-Mining: A Tested Concept for Small Mines   59
*S.L. Chakravorty*

## SECTION 2: ENVIRONMENTAL PERSPECTIVE

Australian Initiatives for Best Practice Environmental Management in Small Scale Mining in the Asia-Pacific   67
*Peter Hancock* and *Stewart Needham*

| | |
|---|---:|
| Integrated Environmental Management in Small Scale Mining — A Bolivian Experience<br>*Guillermo Cortez* | 81 |
| Environmental and Social Considerations during Abandonment of Small Scale Underground Mining Operations: Case Study of a Copper Mine<br>*A. Santha Ram and A.N. Bose* | 83 |
| Development of an Algorithm for Integrated Environmental Management Information System for Small Scale Open Cast Mines of Himalayan Region<br>*A.K. Soni and A. Swarup* | 93 |

## SECTION 3: TECHNICAL DEVELOPMENTS

| | |
|---|---:|
| Building a Better Sluice Box<br>*Don Stewart* | 111 |
| Equipment Selection for Small to Medium Scale Mines<br>*Prabir Paul, G.C. Mishra and D.K. Panda* | 129 |
| Identification of Technological Problems in Small Scale Mining in Tanzania; A Move towards Poverty Alleviation<br>*W. Mutagwaba, A. Mlaki and R. Mwaipopo-Ako* | 141 |
| Birth of Small and Tiny Beneficiation Plant<br>*Shekhar Chakravarty* | 153 |
| Software and Hardware Requirements for Small Scale Mining Sector<br>*A. Santha Ram* | 161 |

## SECTION 4: SHARING EXPERIENCE WORLD WIDE

| | |
|---|---:|
| Prospects and Problems: Small-Scale Gold Mining in Papua New Guinea and the Philippines<br>*Don Stewart* | 173 |
| Small Scale Mining and the Environment in Zimbabwe: The Case of Alluvial Gold Panning and Chromite Mining<br>*Oliver Maponga* | 185 |
| Small and Medium Scale Mining in Indonesia, History and Current Activities — 1996<br>*Antony H. Osman* | 213 |

| | |
|---|---:|
| Marble Mining in Small Scale Sectors in India — Problems, Prospects and Suggestions<br>*A. Bhatnagar* and *S.K. Mukhopadhyay* | 225 |
| Recent Trends of Dimensional Stone Mining in India<br>*K.U.M. Rao* and *A. Bhatnagar* | 235 |
| Artisan Mining of Coal in the Garo Hills, Meghalaya<br>*D.K. Jain* and *A.K. Sural* | 243 |

## SECTION 5: ECONOMIC PERSPECTIVE

| | |
|---|---:|
| Competitive Edge of Minerals from India's Small Scale Mines in Global Market<br>*C.S. Jha* | 255 |
| Economics of Small Scale Resources: A Collective Mining Effort<br>*Suman Banerjee, Sayeri Banerjee, Sajal Dasgupta* and *S.M. Chatterjee* | 269 |

## SECTION 6: MISCELLANEOUS PAPERS

| | |
|---|---:|
| An Insight into Entrepreneur-Employee Relationship in Small-Scale Mining Activities<br>*S.M. Mukhopadhyay* and *J. Bhattacharya* | 275 |
| Modern Education and Training for Small-Medium Scale Mining Mineral Policy Needs<br>*M.B. Katz* | 283 |
| Mining Information System (MINIS) and Global Network<br>*Partha Pratim Chaudhuri* | 289 |

# SECTION 1

# Small and Medium Scale Mining/ Issues Management

# ADDRESSING SOCIAL TENSIONS IN MINING: A FRAMEWORK FOR GREATER COMMUNITY CONSULTATION AND PARTICIPATION

*Mamadou Barry*
Consultant, Industry and Mining Division, Department Energy and Industry
The World Bank, Washington

The launching of a mining project in remote, undeveloped areas usually involves three key stakeholders — the government, the mining company, and the local community — pursuing different and sometimes conflicting objectives. The government expects the project to contribute to the local and national economy. It provides the legal, regulatory and institutional framework for mining development and receives a share of mining revenues in return for access to mineral resources. The mining company brings risk capital and expertise to the project and expects to achieve a rate of return on investment commensurate with the risk undertaken. The local community has a traditional claim to the land and other natural resource inputs needed for mining development and expects to have a say in the planning of the project and to capture a share of the mining benefits. An important subset within the local community is the group of artisanal and small-scale miners whose interests in a mining project can be at odds with those of the government and foreign mining companies.

## Social Tensions in Mining

In the past, foreign mining companies relied on host governments to foster economic linkages and interdependencies between mining operations and the local economy. However, because mining operations are highly capital intensive and technologically complex, they are often detached from the local economy and culture. This lack of direct linkages makes it difficult for benefits to trickle down to the local economy. In addition, the advent of a foreign mining company is often followed by the displacement of small-scale miners and the loss of access to land. Since small-scale and artisanal mining activities provide a critical source of alternative income for many members of the local community, government-imposed restrictions of access to prospective land in favor of large-scale mining can be a significant source of tensions between local small-scale miners and foreign mining companies.

Tensions arise between small-scale miners and international mining companies when local miners begin to perceive modern mining as a foreign activity that appropriates their ancestral lands, exploits their natural resources, disrupts their traditional lifestyles and upsets the ecological balance of their habitat without generating commensurate social and economic benefits for them. In Venezuela for instance, the displacement of artisanal miners in the Guyana region following a 1977 decree restricting small-scale mining to selected areas has left small-scale miners scrambling for new sites. As in other mining countries in Latin America, these restrictions have driven small-scale miners into properties acquired by international mining companies. The same situation exists in Africa where the displacement of small-scale miners in favor of foreign mining companies has led to encroachment of privately-owned properties in Burkina Faso, Guinea, Ghana and Tanzania. Social tensions in mining are not confined to the developing world. In industrialized countries, they often arise from disputes over land claims, sanctity of religious sites, and environmental degradation.

## COMPANY-COMMUNITY PARTNERSHIP INITIATIVES

Concerns over the potential risks of social conflicts have led to a rethinking of the usual policies which by-pass local communities during project preparation and negotiation. In recent years, community consultation and participation have become an integral part of the mining development process. This trend is reflected by corporate initiatives to forge constructive company-community partnerships for mining development as a way of reducing social tensions. International mining companies are increasingly incorporating community concerns into the planning of their operations and are devising appropriate strategies to forge partnerships with the local communities, particularly the small-scale miners. While there is no universal formula for establishing mutually beneficial relationships with small-scale miners, the choice of a community participation strategy must take into account the ownership arrangements of mineral resources and the expectations, values and needs of the community. Some noteworthy initiatives toward strategic company-community alliances are described below.

## Community Consultation

This participatory approach establishes a framework for cooperation to achieve community consensus toward socially, economically and environmentally sustainable mining development. It aims to help companies reflect community values, expectations and needs into the design and implementation of mining projects. The Whitehorse Mining Initiative organized by the Mining Association of Canada in 1992 is an example of company-community partnership for mining development emphasizing community consultation. The initiative involves representatives of the mining industry, public institutions, labor unions, aboriginal peoples and the environmental community. Consultations among these groups began on February 1993 and led to the signature of an Accord in mid-1995. The Accord enunciates a strategic vision for a healthy mining industry that shares opportunities with aboriginal peoples. A participatory approach was also used by BHP Minerals in the mid-1980s to form a community-based consultative group during the planning of the development of the Beenup titanium deposit in Australia.

## Social Activism

This model rests on company support of actions oriented toward improving the living standards of the local community through targeted investments in education, health and agriculture. Battle Mountain Gold uses this method in the village communities near the Kori Kollo gold mine in Bolivia. Through its Inti Raymi Foundation, Battle Mountain Gold has constructed 14 village schools, numerous health clinics and an educational complex including a gym, a computer center and a library. The company has also supported agricultural production through the construction of greenhouses, irrigation ditches, dams and wetlands. Other community activities supported by the Inti Raymi Foundation include the construction of a slaughterhouse to enable natives to directly sell their meat and the establishment of a women-run weaving cooperative which produces hand-made scarves, shawls and blankets for export to European and North American markets. RTZ also supports foundations in its Rossing and Palabora operations in South Africa. In the same vein, Newmont Gold Company directly funds training and scholarship programs in Indonesia and Uzbekistan and school lunch programs in Peru.

## Institutional Cooperation

This approach aims to reduce existing tensions by developing a mutually beneficial relationship with displaced small-scale miners. Placer Dome has favored this model to normalize relationships with artisanal miners at its

Las Cristinas gold property in Venezuela. The company has put forth a number of initiatives to support the local economy and to improve its relationship with artisanal miners who previously operated within its concession. Actions being considered under the institutional cooperation arrangement with artisanal miners include the delineation of areas suitable for artisanal mining, the provision of technical assistance and advice, the support of self-help programs to improve education, water supply and sanitation, and the maximization of purchases of local inputs.

## Leasing and Operating Agreements

These are used in situations where local communities have clearly defined rights over subsurface mineral resources. These arrangements are common in North America and Australia where mineral deposits lie within the settlement lands of indigenous peoples. They enable tribal authorities to enter into an agreement giving the mining company the right to build and operate the mine. In return, the local community receives royalty payments and a share of mining profits. Such an agreement was signed in 1982 between the Northwest Alaska Native Association (NANA) and Cominco for the development of the Red Dog zinc mine in Alaska. The NANA/Cominco agreement provides financial benefits to the natives through royalty payments (4.5% of the net smelter return), profit sharing (25% of the net proceeds after the company recovers its capital investment escalating by 5% each year to a maximum of 50%). In addition, NANA members participate in joint management committees and enjoy preferences in employment and training. By 1993, NANA accounted for 25% of the mine workforce receiving an estimated US$ 9 million in direct and indirect benefits. A similar approach used by BHP in the United States provided the Navajo Nation of New Mexico with royalties estimated at US$ 25 million in 1995.

## Managed Cohabitation

This is a voluntary agreement by which mining companies accept artisanal mining but specify the conditions under which it can continue within their concessions. Golden Shamrock uses this model in its gold properties in Ghana and Guinea. The company allows artisanal gold mining to continue in the shallow alluvial resources within its concession provided that company and government requirements are fully met. In Guinea, the company holds a large lease which contains alluvial gold resources estimated at 450,000 ounces but has opted to pursue primary gold as its main target. Its current policy is to allow artisanal mining of alluvial resources to continue as long as it does not interfere with the company's hard rock mining operations. A variation of this model involves working out a satisfactory marketing arrangement with small-scale miners operating within a company's concession. This approach has been recently adopted by Howe Centrafrique to resolve tensions with artisanal diamond miners in the Central African Republic and is being contemplated by Placer Dome in the Las Cristinas gold project in Venezuela.

## THE ROAD AHEAD

While local communities expect and need to receive a fair share of the economic benefits of mining, no benefits will be realized unless mining companies are able to operate in a stable and cost-competitive environment. This requires as a first step the establishment of an orderly and safe environment for the conduct of small-scale mining. The World Bank is of the view that a comprehensive approach, rather than a piecemeal solution, is needed to address the legal, technical and financial constraints which affect the development of small-scale mining. Such an approach would focus on establishing the enabling legal and institutional arrangements for small-scale mining, alleviating technical and financial constraints and improving environmental, living and working conditions in mining sites.

Land conflicts can be reduced substantially by giving legal recognition to small-scale mining with all the rights of transferability and mortgageability of mining titles. To this end, governments will need to streamline registration and licensing procedures, liberalize marketing arrangements and strengthen small-scale mining institutions. But governments alone cannot effectively resolve the social tensions in mining. Rather, the solution to social tensions will require a constructive partnership to channel the actions of government agencies, community leaders, miners' associations, NGOs, international donor agencies and international mining companies.

The World Bank welcomes such a partnership and stands ready to play a catalytic role by assisting governments to establish enabling conditions for the development of small-scale mining, by collecting and disseminating best practices, by financing policy reforms and targeted actions such as microfinance programs. Bank-financed activities in small-scale mining have been gradually operationalized in several countries, with projects under preparation or currently underway in Ecuador, the Andean countries of Peru, Bolivia and Chile and a number of African countries including Burkina Faso, Ghana, Guinea, Mali, and Tanzania.

Reducing social tensions also requires a change of attitude from both foreign mining companies and small-scale mining communities. International mining companies should recognize that it is in their best interest to forge mutually beneficial relationships with small-scale miners and to be responsive to the development objectives of the local community. Over the long run, this can substantially reduce the risks of social tensions and possible disruption of operations. In return, local communities should adjust their expectations and help foster a business-friendly, stable and competitive environment for mining operations. These shifts in attitudes and the corporate-community alliances they help bring about are crucial for the development of a strong, socially responsible and environmentally sustainable mining industry.

# SMALL-SCALE MINING: TIME FOR DEEDS NOT WORDS

*Norman S. Jennings*
Senior Industrial Specialist, International Labour Office

**INTRODUCTION**

This brief paper — on the occasion of another gathering of those involved in, or on the fringe of, small-scale mining activity — questions the need for and direction of activity by agencies such as ILO and NGOs, as well as mining companies and governments, on behalf of those generally absent from such conferences, the small-scale miners themselves. The paper concludes that, even when small-scale mining can operate by itself — which it does for most of the time — such operations might not be for the best for all concerned. Appropriate, integrated external influence or involvement can help small-scale mining realize its full potential — for the mine and its owners and workers, for the environment and for the exchequer. Such involvement needs to be carefully developed and implemented — in consultation with all concerned. The lack of hard information on small-scale mining is starting to become less of a problem as information and data from different activities filter into newsletters and journals; but it is still sporadic. The revamping of Small Mining International (SMI) with a new outlook and approach should enable it to become the focal point and clearing house for information on all aspects of small-scale mining. In this way, news of successful initiatives can be disseminated and, hopefully, result in their duplication. In view of the very real occupational safety and health risks that are prevalent in many small-scale mining operations, the paper addresses the need to incorporate occupational safety and health considerations in all small-scale mining programmes. The paper highlights the role of the new Convention on Safety and Health in Mines that was adopted in 1995 as a means to establish minimum safety standards for all mines.

## SMALL-SCALE MINING: SUCCESS OR FAILURE?

Small-scale mining falls into two broad categories. The first is the mining and quarrying of industrial minerals and construction materials on a small scale. These operations are mostly for local markets and they exist in every country. Regulations to control and tax them are often in place. Informal or illegal operations at this level are generally due to a lack of inspection and the lax enforcement of regulations rather than to the lack of a legal framework, much the same as for small manufacturing plants. The second category is the mining of relatively high-value minerals, notably gold and precious stones. The output is generally exported, through sales to approved agencies or through smuggling. The size and character of small-scale mining of this type has made what laws there are impossible to apply or inadequate.

The "success" of small-scale mining depends on how you look at it. Small-scale mining provides employment, income and foreign exchange; it enables otherwise non-economic resources to be developed; it helps to avoid rural-urban migration; it has led to large-scale mining; and it maintains the link between people and the land. From another point of view, small-scale mining can be considered exploitative — of people, land, resources and the environment; unhealthy, dangerous, uncontrolled or illegal; more trouble than it is worth. Which of these is true? They both are, to a certain extent. The social and economic complexity of small-scale mining and the fact there is no model on which to develop a sound theory or programme mean that, despite the conferences, the guidelines, and the programmes of action, not much has changed.

At the macro-level, small-scale mining has rarely achieved its full potential. The much vaunted ability of small-scale miners to find deposits, particularly of high value minerals is one thing. The knowledge, ability and will to exploit them fully are another. Output and productivity from small-scale mines are lower than they could be. Returns to the economy are certainly less than if effective tax, purchase, pricing and foreign exchange regimes were in place and implemented. Unattractive or unworkable schemes lead quickly to illegality and the smuggling of the output to more favourable areas. In Uganda, for example, where gold is mined by artisanal miners, official exports in 1995 amounted to 20 kg compared with 1.3 kg in 1994. The reason? The Central Bank's monopoly on purchases was abolished and licensed dealers were allowed to sell gold directly on the world market. Cash flows from small-scale mining operations quickly when compared with large operations. Also, most of the surplus is generally paid out locally in profit-sharing and wages to a large workforce, resulting in rapid improvements in the local economy, rather than in dividends that are often repatriated and wages to a small workforce, sometimes with a high proportion of expatriates. At the micro-level, some mining company operations have coexisted with small-scale mining, giving them some assistance and thereby having some influence on their activities. Others have sought to stamp small-scale mining that might affect them. NGOs have worked hard and effectively at the local level to introduce appropriate technologies to improve efficiency and mitigate the environmental and health impact of small-scale mining. Some IGOs have undertaken studies and developed guidelines and programmes of action. These include child labour, taxation and land title reform, environmental impact and the role of women and indigenous people. So far they appear to have had little discernible effect. It should be noted, however, that IGOs need the active support and participation of governments in order to develop and implement programmes in this or other areas of activity.

The ILO, for example, as part of its International Programme for the Elimination of Child Labour (IPEC), is undertaking projects in Colombia and Peru to provide educational opportunities and alternative income-generating prospects to remove children from coal and gold mines in three regions in these countries. This work is being coordinated with the international mineworkers' union (ICEM) and with local NGOs and government agencies. At the government level, it was recently reported that the South Asian Association for Regional Cooperation had agreed at ministerial level to ".... end all child labour in hazardous professions by the end of the century." This will be welcome news to the many children who work in small-scale mining, despite their participation already being illegal in most countries. For the most part, there seems to be little interest among small-scale miners in using cheap, readily available and effective technology, such as retorts to capture mercury. There is often no economic incentive for them to do so, since the cost of mercury is not a constraint. Moreover, particularly in the case of itinerant miners, there is frequently no long-term interest in preserving the land for use after mining has ceased.

## "ASSISTANCE" — WHO DOES WHAT?

Governments have intervened in small-scale mining when they have had to — to improve revenue flows, on behalf of indigenous people, and at the request of mining companies. There are, regrettably, few examples of specific policies for small-scale mining that cover its development and operation. Specific legislation is also rare, and many small-scale mines fall outside the ambit of traditional mining laws. The overwhelming problem facing governments in attempting to regulate small-scale mining is the lack of a trained inspectorate. The result has been an inability to introduce controls at all, or the failure to follow-up or monitor mining operations so that problems have not been dealt with at an early stage and projects have foundered. The result has been the continuation of unsatisfactory practices.

Our task should be to show small-scale miners that, without unduly constraining their activities, there might be a better way to go about their mining. Better for them in terms of health and wealth; better for the land; and better for the country. Unless small-scale miners can be convinced that the involvement of well-meaning agencies can produce immediate tangible benefits, we might as well forget it. How can this be achieved? Probably a little at a time, using good successful examples to convince people that there is a way for them to be better off without jeopardizing resources, land and people. Agencies like SMI can play an important role in collecting and disseminating information from and to the grass-roots level. National or regional offshoots of SMI should be encouraged; they can get out among the miners themselves, convince governments and encourage mining companies. On the ground, we should use existing organizations, not muscle in on top of them. Existing organizations (such as mining associations or NGOs) work quickly, are less bureaucratic, know the local situation and have a rapport with the people involved.

There is a framework for the development of small-scale mining. The "Harare Guidelines" provide guidance for governments and for development agencies in tackling the different issues in a complete and coordinated way. There is no point in looking at one aspect of small-scale mining without addressing the others; the linkages are so tight. Last year in Washington, the question of artisanal mining was addressed at the initiative of the World Bank. The outcome was a draft strategy to move artisanal mining gradually into the formal sector. This strategy underscored the need for partnership between the key players that has been highlighted on several previous occasions. Setting guidelines is the easy part. Finding the will and the resources to implement them is proving more difficult.

With the recognition by the World Bank that much small-scale mining activity is linked to poverty, which is at the forefront of its programmes, its formidable resources might be brought to bear in a constructive way to the benefit of the millions involved. A sensitive integrated approach is required.

In recognition of the many labour and social problems that beset small-scale mines, and of the need to exchange experiences of good practice in small-scale mining, the ILO had planned to hold a meeting in 1997 of governments and employers and workers' organizations to address a range of issues. This meeting was cancelled because of budgetary problems and it remains to be seen whether it will be re-selected for the biennium 1998–99. A decision on this will have been taken by the time of this Conference. By discussing and clarifying the role of governments, the social partners and the ILO we hope to be able to raise the profile of small-scale mining within ILO and assist in providing the means for small-scale mining to ensure safe and productive employment and good working conditions. These will inevitably contribute to the achievement of higher productivity, better resource management and a lessening of environmental impact.

## SAFETY AND HEALTH

Mining is often a hazardous industry, more so than most. World-wide, the formal mining sector employs only 1 per cent of the workforce, but it accounts for almost 8 per cent of fatal accidents at the workplace. The acknowledged under reporting of accidents makes the true toll likely to be even higher, particularly when one takes into account the fact that, in many countries, the number of people working in small-scale informal or under-regulated mines exceeds considerably the workforce in the larger-scale formal mining sector. Much of this employment is in rural areas where it is important in alleviating underemployment. Unfortunately, however, many of these jobs are precarious and are far from conforming with international and national labour standards. For example, child labour is not unknown in small-scale mines, hence ILO activity through its IPEC programme. Moreover, where there are statistics, they tend to show that the rate of occurrence of fatal accidents in small-scale mines is routinely six or seven times higher than in larger operations, even in industrialized countries. That is not to say that there are no safe, clean small-scale mines — there are, but they tend to be in the minority. The situation concerning non-fatal accidents and occupational diseases is even less certain because of the almost total lack of reliable data. Anecdotal "evidence" points to a sorry state of occupational health and safety in small-scale mines — at the workings themselves and also in the neighbouring living and farming areas.

The geographically scattered nature of small-scale mining activity, its rapid rise during ôgold rushesö, lack of resources in mines inspectorates and a desire to avoid drawing attention to illegal or quasi-legal mining have perpetuated the lack of information on what is increasingly an activity of considerable economic importance. The fact that small-scale mines are generally outside the scope of activities of employers' and workers' organizations — which in the large, formal sector have rights, obligations and an influence on mining operations — means that the State has had to shoulder all the responsibilities for managing small-scale mining. It is rarely equipped for this task and so, in many cases, it has not been able to do so. Hence the need for external assistance to bolster the size and technical competence of mining inspectorates and to move small-scale mining into the formal sector. The ILO will assist wherever it can in this regard. Mining companies will have a role too.

## CONVENTION 176

A new *Convention on Safety and Health in Mines* and an accompanying Recommendation were adopted by the International Labour Conference in 1995 after two years of intense discussion. The Convention, when ratified, has the force of law. The Recommendation sets out in detail an internationally agreed benchmark to guide national law and practice.

The Convention covers all mines. While there are provisions for excluding certain categories of mines — provided the overall protection afforded at these mines is not inferior to that which would result from the application of the Convention — plans for progressively covering excluded mines must be made. Since mines that are considered to be illegal would presumably be excluded for all time from the coverage of the Convention, the importance of bringing small-scale mines progressively into the formal sector increases.

The Convention specifies procedures for reporting and investigating mine disasters, accidents and dangerous occurrences, and the publication of relevant statistics. It covers employers' responsibilities to eliminate or minimize risks to the safety and health of workers in mines, as well as addressing the rights and duties of mineworkers in relation to safety and health. It refers to the establishment of inspection services to ensure effective enforcement of its provisions.

It should be stressed that it is not necessary for a country to ratify the Convention for it to put its provisions into effect. The provisions of the Convention are already being included in collective agreements in several countries. Also the Convention sets minimum standards and there are many mining countries which already have safety and health regulations that exceed the provisions of the Convention. The objective is to enable those countries that do not have adequate legislation on safety and health in mines to achieve an agreed minimum level and, in ratifying and implementing the Convention, to be included in the ILO's supervisory procedures.

International and national mining trade unions are mounting an unprecedented effort to obtain quick and widespread ratification of the Convention. They also recognize that, because many small-scale mineworkers are not represented by trade unions, they have a special role in helping these mineworkers, many of whom work under very poor and dangerous conditions.

An important issue for many developing countries will be the strengthening of their "competent authorities" in order to ensure that the Convention's provisions are being followed. With infrastructure geared towards the well-developed formal mining sector, the sudden influx of small-scale mining activity would stretch the most efficient authorities, let alone those that are already poorly staffed and funded. External assistance may be required to bolster the size and technical competence of mining inspectorates. The ILO, in conjunction, where appropriate, with other agencies, will assist in this regard.

The Convention provides a floor—the minimum safety requirement against which all changes to mine operations should be measured. The need for continued vigilance remains paramount during the substantial changes that are taking place in mining world-wide. Positive results should be examined and, where appropriate, replicated through the concerted efforts of all concerned parties. This implies a higher degree of information exchange and collaboration than has been the norm. The ILO and SMI could facilitate the collection and dissemination of examples of "good practice" in improving mine safety.

It is hoped that the new mining Convention that has set the principles for national action on the improvement of working conditions of the mining industry will be widely ratified and will lead to significant and lasting improvements in safety in mines large and small.

## CONCLUSIONS

The time has come for us to put up or shut up as far as small-scale mining is concerned. For me, shutting up is not an option. Small-scale mining is more often than not considered as a "fringe" activity that is always the responsibility of someone else. But an economic activity that employs over 6 million world-wide is hardly on the fringe. Also, small-scale mining is generally closely associated with other mining. Despite this, it is treated differently or ignored. This isolation has suited all concerned at one stage or another. However, when developing countries are struggling to extract the maximum benefit from their mineral resources, small-scale mining needs to given full recognition for its contribution to national wealth and employment and receive proper consideration when policies and regulations are developed and implemented. It is not necessary, in my view, to attempt to force small-scale mining into a rigid mould. This would destroy it. But it should be drawn under a mining "umbrella", not left out in the cold. For this to occur, there needs to be a commitment of resources to change the perception of small-scale mining, from within and without. Also, governments and larger mining companies need to have the will and resources to ensure that they can assist small-scale mining to have a productive and legitimate existence. The onus is on governments to move consideration of small-scale mining higher up their political and economic agendas. Only when this occurs will sustainable progress be made. Agencies such as ILO are in a position to assist, but the drive must come from within.

## REFERENCES

*Mining Journal* (London), 23 Aug. 1996, p. 142.

Burke, Gill: "Policies for small-scale mining: the need for integration", in *Conference proceedings: Mining and mineral resource policy issues in Asia-Pacific: prospects for the 21st century*; Australian National University, Canberra; 103 Nov. 1995, pp. 103–106.

*Financial Times* (London), 23 Aug. 1996, p. 8.

United Nations Interregional Seminar on Guidelines for the Development of Small/Medium Scale Mining; Harare; 15–19 Feb. 1993.

*Proceedings of the International Roundtable on Artisanal Mining*; World Bank, Washington, DC; 17–19 May 1995

World Bank: *A comprehensive strategy toward artisanal mining*; Industry and Mining Division Draft Paper; Aug. 1995 Washington, DC.

# STRATEGIES FOR DEVELOPMENT OF SMALL/MEDIUM SCALE MINES IN AFRICA

*Pierre A. Traoré*
Officer-in-Charge, Natural Resources Division
Economic Commission for Africa, Addis Ababa Ethiopia

## INTRODUCTION

The African continent has a long small/medium scale mining tradition as far as this term applies to any mining operation depicted by the simplicity of the equipment and technology used, and the low level of investment. Small scale mining has formed the basis of African minerals production for centuries. However, there were a long period of decline and, in its modern mining history, small/medium scale mining activities gathered momentum especially during the past 15–20 years.

## CURRENT STATUS

More than 40 minerals — including precious, semi-precious, heavy, industrial minerals and construction materials — are currently extracted by small/medium scale mining in almost all African countries. In many countries — i.e. Burundi, Central African Republic, Chad, Congo, Mozambique, Rwanda, Uganda — all the mineral production results from this type of mining. In the other countries, including some with large scale mines, such as Guinea, Mali, Namibia, Tanzania, Zimbabwe, small scale mining production still plays a significant role.

The magnitude of small/medium mining in Africa is also shown by the number of miners involved in the activities. It is estimated that more than one million people are directly involved in. The number becomes far more if the related activities — blacksmiths, traders, caterers etc. — are taken into consideration.

As a result, small/medium operations have various impacts among which economic, social and environmental ones, are indeed worth of interest. Therefore, in this paper the current status of small/medium scale mining in Africa is considered under three corners, namely (a) the economic, social and environmental impacts; (b) the constraints which African women are facing; and (c) the actions taken by African Governments for a sustainable development of the sector.

## Predominant Impacts of Small/Medium Scale Mining in Africa

*Economic and social impacts*

The economic and social impacts of small/medium scale mining activities are tangible both at the local and national level. Besides, they concern individuals, families and communities directly or indirectly involved in these activities.

*As regards the economic impacts*, in rural areas, small/medium scale mining confer to poverty alleviation: (i) it constitutes a response to the crucial problem of survival of a large number of poor people; (ii) it contributes to the improvement in the conditions of life of these people; and (iii) it stimulates the trade and economic activities — in particular when gold and other precious minerals are concerned. At the national level, the earnings from small/medium scale mining offset the deficit of the balance of trade and payment. For example in Burkina Faso, the exports revenues from gold produced by artisanal and small/medium mining accounted for about 14% of the overall export earnings of the country, in the earlier 1990s.

*Concerning the social impacts*, they are both positive and negative. This results from the fact that on the sites of activities—in particular for precious minerals—the movement of persons is very intense and the individuals come from different backgrounds with varied objectives and motivations. Some are interested in engaging in mining and/or trading activities while others are attracted by unhealthy deeds (theft, vagrancy, brigandry, etc..). As a result, the most important and glaring impacts are: (i) creation of jobs opportunities and minimization of the rural exodus; (ii) establishment of some infrastructures; (iii) use of narcotic drugs; (iv) inculcation of degrading habits; and (v) persistence of conflicts and tensions.

In a social framework characterized by a weak industrial fabric and marked by chronic under-employment which reaches unequalled proportions, the small/medium mining activities are a valuable source of employment opportunities. In fact, even if some call this employment "informal", it is all the same true that the number of people involved in it is relatively high. More than 1,000,000 small miners are reported in Africa. If we add all those who are engaged in related service activities — blacksmiths, traders, caterers, etc.— this figure becomes surely higher.

With their income increasing, the actors in the African small/medium scale mining have the possibilities to contribute to the establishment or to strengthen various socio-economic infrastructures, the most common being schools, health care centres, maternity centres and mosques. They also contribute to the improvement of the communication facilities between the surrounding agglomerations and the activities areas by financing construction and maintenance of roads and bridges like in Burkina Faso, Côte d'Ivoire, Guinea, Mali, Zimbabwe.

Unfortunately, the fever for a speedy accumulation of money from mineral gain motivates the miners to work as long as possible without feeling tired. To do so, they are led to taking excessively narcotic drugs of all types. In the long run, the consumption of such substances has some adverse effects: mental depression and imbalance are frequent among the consumers who become a real danger for the all community at the extraction sites. In addition, the very high concentration of individuals from all origins and with different morality in the activity areas, encourages the development of degrading habits. Thus, theft, delinquency, banditism and prostitution settle in and make the sites the home of insecurity and sexually transmitted and contagious diseases.

Finally, coexistence on the sites of people with divergent interests often ends up with clash between individuals or groups of individuals. In particular, most of the conflicts — notably in the case of gold — take place between the gold washers during excavation activities. The root causes are either belief in irrational practices or conflict of interests. It often happens that owners of pits accuse each other of substituting mineralized veins and that out of sheer jealousy or to prevent the other from reaching it. It can also be simple theft of ones mined materials. Another source of conflict is sometimes the clash between the gold washers, the mining permit holders and chiefs of customary law.

### Environmental Impact

Since a decade, artisanal and small scale mining activities have increased considerably in Africa, mainly regarding precious and semi-precious minerals extractions. For example, in the Liptako-Gourma region in West Africa — which includes Burkina Faso, Mali and Niger — the small-scale gold mining has largely spread since 1984 with high concentration of people of 5,000 to 10,000 individuals on a single site. This entails high physical pressure on the natural environment.

As a matter of fact, in this rush for gold, the mining is carried out in an anarchic manner and without any concern about the rehabilitation of the areas mined. In addition, excessive cutting and felling down of trees for the cribbing of pits and for dwelling and energy (fuel wood) are made. These largely uncontrolled activities result inexorably to: a massive and gradual destruction of the ecosystem; an accelerated deforestation; and a ravage of the plant cover.

Furthermore, the chaotic mining of pits and galleries across the sites, without complying with the rules of the art, and the accumulation of huge quantities of sterile excavated materials lead to the degradation of soil. The mines, excavated without any frame, remain there permanently and thus constitute a mortal danger for the people living on the sites because of the frequent caving in and landslides they cause.

The recent introduction and use of chemical products by some small miners — i.e. mercury — for the processing and production of fine gold constitutes a permanent risk of pollution of the rare water resources and the surrounding of the region.

The high concentration of small miners at the operations sites and the huge food requirements of these people compel many of them to carry out farming during the rainy season either on the sites or their immediate neighbourhood. There is consequently an anarchic occupation of land and a gradual destruction of the thin layer of arable land.

## Actions Undertaken by Some African Countries to Increase Positive Impacts and Minimize Negative Effects of Small/Medium Scale Mining

At national and sub-regional levels, many African governments are in the process of establishing new strategies aim at: (a) reinforcing the positive impacts of the small/medium mining and (b) minimizing their negative effects both in social and environmental domains.

A study carried out by the Economic Commission for Africa in 1994–1995 on the status of the African mining sector demonstrated that various measures were underway. They include: (i) the legalization of and an attempt to clarify the status of small/medium mining; (ii) an increased attention to the environmental concern; (iii) the establishment of specific institutional and administrative framework; (iv) the reinforcement of technical assistance; (v) changes in the commercialization pattern; and (vi) the reinforcement of sub-regional cooperation.

By the end of 1995 about 36 countries had legalized such operations or were in the process of doing so. Besides, several countries had established — or were establishing — specific administrative and/or technical institutions with a view to providing technical assistance to the small miners. Additionally, as regard the commercialization of mineral products, there were a general trend towards the liberalization based on a free market regulation. Finally, a significant interest in cooperation for the development of mineral resources — including in small/medium scale mines operations, were noticed in particular in the southern and western Africa with the impulsion of the Southern African Development Community — SADC- and the Authority for the Integrated development of the Liptako region — ALG — respectively.

To alleviate the negative social effects of small/medium scale mining, accent is also put on the provision of extraction sites with security and health facilities. Some countries — like Niger — have established specialized units to ensure security, crack down on the users of drugs and to enforce measures aimed at minimizing smuggling. In connection with health care the facilities established, sometimes with the contribution of the various operators, provide first aid — like in Burkina Faso and Mali.

Concerning the environment, there is a gradual awareness of environmental degradation created by mining industry in general and by artisanal, small/medium mining in particular. Thus, provisions of legal acts related to the environmental rehabilitation are provided and environmental impact assessments have now become a requirement in most mining regulations. In this respect, some countries have established a rehabilitation or environmental restoration fund sustained either by an environmental tax incorporated in the charges for mining rights (i.e. Guinea) or by the levy on royalties and benefits paid by mining companies (i.e. Ghana).

## Constraints Faced by African Women in Small/Medium Scale Mining

While different efforts are made to improve the performance of the small/medium mining and to reduce its negative effects, the gender issues still remain. Despite their great number — in Guinea and Mali about one out of three small miners is a woman — women continue to face technical and socio-cultural constraints, including the difficulties to access bank credits.

One of the greatest difficulties for women is the insufficient knowledge of mining techniques. As the general level of schooling among girls is far below that of boys and because this situation has worsened over the past three decades, a high proportion of illiterate women among the female population are involved in small/medium scale mining. Therefore, women are far more handicapped in learning the techniques needed for good mining.

The second constraint has to do with the difficulty of access to bank loans needed to purchase mining equipment and machinery. Indeed, socio-cultural outlooks in most of the African countries establish the man as the head of the family. His consent is needed for his wife to qualify for a bank loan. As it happens, women often bear the cost for the daily survival, particularly of children in the family. Therefore, in the absence of financial assistance from the bank, they find themselves compelled to devote all the meagre resources they have earned from informal activities to that purpose, instead of acquiring mining equipment.

Another difficulty which tends to reduce the full participation of women in small/medium scale mining has to do with some unprogressive traditional ideas. For example, in countries like Ghana, women are forbidden access to the sites where precious minerals are being mined during their menstrual period. Consequently, in addition to the periods of unforeseen sickness when a woman has to stop work, she is also compelled to suspend mining activities at least once a week each month.

One last problem that women have to face and which might be ascribed to old society habits is the unpreparedness of men to obey women. Many backward people continue to think that women are made to obey but not to command. Therefore, unconsciously, they sometimes hold up the work of women detaining licenses or permits. This has been observed in Zimbabwe where about 300 women have mining licenses.

In conclusion, there are many reasons—not always objective and/or rational—which keep women from participating fully and beneficially in small-scale mining. The loss is to the mineral production of society as a whole, especially with regard to small-scale mining.

## STRATEGIES FOR SUSTAINABLE DEVELOPMENT OF SMALL/ MEDIUM SCALE MINES IN AFRICA

Small/medium scale mining in Africa correspond so far to artisanal mining with the exception of Zimbabwe. The development strategy should address the problem of moving from artisanal operations toward real small/medium scale mines. This includes policy, legal, social and environmental aspects as well as women full participation.

### Measures for the Establishment of Comprehensive Policy and Legal Frame for the Development of Small/Medium Scale Mines

First, the policy should aim at integrating small/medium-scale mining operations into a global national mining development strategy. Such integration should give the same importance to all mineral products, allocating resources on the basis of socio-economic role and not only that of financial considerations.

Second, the legal framework must specify the various level of mining —artisanal, small, medium and large mines — and clearly indicate the permissible ways to move from one type to another. In addition, the legislation provisions should not focus only, as usual on: (i) conditions for the award of mining concessions; (ii) duration; (iii) the covering area over which they apply; (iv) the taxation system and level; (v) the description of the marketing area and conditions. It should also legalize all the types of mines—including artisanal, based on official licenses delivery with incentives measures and legal mechanisms for progressive moving from artisanal to small then medium scale mines.

Lastly, the above-mentioned measures must be reinforced by: (i) establishing mechanisms for training manpower; (ii) reinforcing technical and financial assistance; (iii) facilitating the involvement of national engineers and mining professionals as operators; and (iv) insuring availability and access to appropriate mining equipment and materials.

## Measures for the Improvement of Socio-economic Conditions

The social measures earlier mentioned as being taken by Governments need to be reinforced. In addition to health facilities, effort should be made to provide the sites under operation with infrastructures especially water and electricity. Moreover, safety conditions in mining must be given high attention.

## Measures Aim at Environmental Protection

As for environmental restoration, the existing measures can be reinforced by the imposition of an environmental rehabilitation guarantee fund. This will require any mining operator to contribute to such fund at the time of the issuance of mining permit or authorization. The contribution will be reinstituted at the end of the mining operations, if the miner restores the site in conformity with the provisions of the law governing extraction sites restoration. If the miner fails to carry out the rehabilitation, the Administration which have and may use the right to prosecute the defaulting miner, could defray part of the restoration charges against his contribution to the established fund.

Furthermore, as a set-off against the contribution of their mining activities towards trees and soil destruction, systematize afforestation campaigns on and around small/medium mines areas must be undertaken with the participation of small miners and other relative actors.

## Measures to Address Gender Issues

Actions advocating the full participation of women must result in concrete measures to enable them a full participation. In particular, inadequate technical know-how and difficulty of access to bank credit — which constitute the main evident constraints, should be addressed.

As regards technical aspects, the establishment of training workshops and seminars tailored women working or willing to work in that sector may provide a partial solution to the problem. To that end, such meetings shall be organized not on an ad-hoc basis but in a systematize manner once a year at least. For these seminars and workshops to be successful, they must address unique and specific themes for each session such as: keeping files on grant of permits or credits; mining and processing of ore; marketing; management.

Among the measures to be taken to facilitate women miners' access to bank credits, are: (i) the abrogation by the State of all discriminatory provisions in the legal texts that hold women down as the economic dependents of men. Thus, any requirement for a woman to secure the authorization or even the agreement of her husband to qualify for bank credit must be abolished and banned; (ii) the introduction of a "small feminine mine" window in all bank structure and mechanisms for the assistance to small and medium enterprises on the one hand and those set up to foster the promotion of woman/entrepreneur on the other hand.

Besides, to officially back the women and get them more deeply involved in these activities, the Government shall ensure that every project related to that sector contain a gender component. In that same spirit the State must adhere to the principle of a quota—considered as the minimum to be fulfilled as far as possible and which can obviously be exceeded—between men and women regarding the acquisition of mining permits and authorizations.

Lastly, as regards socio-cultural constraints, sustained sensitization campaigns must be initiated to denounce those constraints that are not founded on any rational basis. Those campaigns must be organized both at national and regional level.

## CONCLUSION

In its current status, African small/medium scale mining are already significant in terms of economic, social and environmental impacts. However, because they are indeed reduced to artisanal operations, their positive economic and social impact are still limited while their adverse social and environmental impacts remain important. To reverse this situation, efforts should be made to establish a structural change by moving from the existing archaic and informal artisanal mining to a rational and entrepreneurial small/medium scale mining.

To achieve this objective, national efforts should be reinforced by the strengthening of sub-regional and regional cooperation. International support will also be needed in the process of the implementation of various guidelines and resolutions related to the development/strengthening of small/medium scale mining in general and to its development/strengthening in Africa, in particular.

# AN INTEGRATED PARTICIPATORY FRAMEWORK FOR THE EXPLOITATION OF MINERAL RESOURCES IN THE CONTEXT OF SUSTAINABLE DEVELOPMENT
## THE CASE OF NAMAQUALAND, SOUTH AFRICA

*Olle Östensson*
United Nations Conference on Trade and Development (UNCTAD)[1]

## BACKGROUND

Mineral resources can generate substantial benefits for the countries in which they are located, as well as for the mining companies which exploit these resources. Exploration and production activities provide revenue to the government in the form of taxes and license fees. Mineral exports generate the foreign exchange needed to purchase imports and finance economic development. Employment opportunities arise for the local population, both in mining and related industries, with concomitant benefits by way of increased incomes and better health and education conditions. New infrastructure and services established for mineral exploitation benefit local communities within and beyond the mining area. Thus, a number of stakeholders — the mining company, central government, local government and local communities — have an interest in mineral wealth.

Small-scale mining, even at the level of artisanal activities, also makes an important contribution to employment and local wealth creation. In many cases, income from small-scale mining supplements household income from other activities and provides a much needed source of cash income. This in turn increases local demand and stimulates other economic activities. In addition, small-scale mining often contributes to raising the level of technical know-how in a community and may form a basis for other small-scale industrial activities.

---

[1] The views expressed in the paper are those of the author and do not necessarily represent the views of the UNCTAD secretariat.

But mineral production also raises problematic issues for sustainable development. In the case of small-scale mining, the problems may be exacerbated as a consequence of the large number of individuals engaged in the activity and its uncertain legal status. "Gold rush" type phenomena have in many cases taken governments by surprise, often occurring in countries or regions with no previous history of mining and without specific provisions regulating artisanal mining in the mining legislation. The miners have often encroached on mining rights held by formal mining companies as well as on land held by other land users, and this has sometimes led to conflicts and violence. With their usually primitive techniques, artisanal miners are unable to extract low-grade ores. Consequently, they may "high-grade" a deposit, making commercial mining impossible. In countries with overvalued currencies and foreign exchange regulations, smuggling of gold and gemstones lead to loss of foreign exchange (and usually to loss of potential income for the miners). The lack of legal status for the artisanal miners has usually also led to destructive environmental practices, including siltation of rivers and pollution from mercury used to extract gold, and hazards to the health and safety of workers. Finally, the lack of government authority in mining areas has led to high crime rates and deplorable social and health conditions.

Even when the number of individuals active in small-scale mining is relatively low and the miners are drawn from the local population, serious problems may nevertheless result. The mining can entail disruptions to the traditional production activities, the life style and the culture of local populations. It may induce suddenly increased demands on other scarce natural resources, such as energy and water, and result in damage to ecosystems. Moreover, individual mineral deposits are finite: when exploitation ceases local communities are forced to find alternative occupations and livelihoods; and this in turn can lead to further social dislocation and increased pressure on the natural environment.

These problems tend to become exacerbated in a developing country context where the institutions, techniques and skills needed to deal with them may be lacking or deficient. However, local and national government bear the responsibilities for resolving the long-term, socio-economic issues facing the populations over which they have jurisdiction. Agenda 21 sets out the broad lines along which governments have undertaken to act in this regard.

The issues surrounding the exploitation of mineral resources have so far received less attention in the follow-up to the United Nations Conference on Environment and Development, as compared with energy, soil and water resources. Nevertheless, as noted above, important sustainable development issues arise when mineral deposits are exploited for commodity production. The issues, which can have far-reaching implications for land use, are varied and complex. They span various disciplines in the social and natural sciences, as well as long time periods—up to fifty years and beyond. The task of exploring development alternatives, and eventually of making decisions, is further complicated by the number and variety of interests and institutions involved and, often, by a lack of channels for constructive communication. Resolving these issues calls for an integrated and comprehensive approach within a well-managed framework.

UNCTAD has recently initiated a project which is designed to make a concrete contribution to the implementation of activities set out in Agenda 21 for the sustainable and integrated management of land resources. The project aims to enhance the capacity of public and private-sector agents of developing countries to address the complexity of sustainable development issues which arise when mineral resources are exploited for commodity production. Specifically, it is intended to:

- Devise a *model framework* to assist in the identification, analysis and management of long-term, inter-disciplinary issues;
- Promote the use of this framework in individual developing countries, with the support of national and regional institutions; and
- Provide *training* on tools, techniques and methodologies to support a cooperative approach to sustainable development involving all the stakeholders and including local communities.

While the approach described below will not resolve all the problems associated with mining in general or with small-scale mining in particular, it may help to identify and clarify the issues involved, and it could serve as a powerful tool for policy evaluation, implementation and monitoring.

## PROJECT CONTENT

### Model Framework

A comprehensive approach to the study and management of a mineral resource event — such as widespread small-scale mining — requires a framework comprising at least the following two components:

- A geomanagement system (GMS) — essentially a geo-referenced management tool of the natural resource location — designed to facilitate the analysis of inputs from a variety of sciences, to situate the habitats of local communities, and to visualize physical and socio-economic change over time, both past and projected.
- An economic/ecological model — specifying the commodity production processes and their interaction with the larger economy, including their impact on environmental resources. This model, which relies on

techniques such as input/output analysis, regional economics, natural resource accounting and dynamic simulation methods, is used to visualize the long-term feedback effects of natural resource projects.

These two components need to be supported by financial information — specifying how the commodity/natural resource project is to be financed and executed — as well as information on the different institutions and interests involved and how they are to communicate with each other.

## Training

The training programme focuses on strengthening capacities for a participatory process in decision-making in order to provide incentives for local initiatives and enhance local management capacity. The strategy envisioned is to focus on specific problems in mineral resources management and to capitalize on the existing databases, maps, and environmental and socio-economic studies that are available.

The specific training objectives are:

1. To provide the necessary tools, techniques and methods that will assist local and national agencies in the assessment of environmental and socio-economic impacts of alternative strategies of resource management, and enable other interests, including, in particular, local communities, to formulate their concerns in a way that can be integrated into the decision-making process.
2. To provide training in the application of management tools (dynamic geographic information systems, decision-support systems, diagnostic tools) that can be used by all the stakeholders, thus ensuring a common frame of reference which will promote constructive dialogue and search for solutions acceptable to all interests.

An integral aspect of the training is its emphasis on hands-on experience using personal computers and adaptable, user-friendly software programmes. In order to increase the acceptability of the methods taught and utilized, emphasis will be given to training of trainers who will in their turn be responsible for the training given to others. These trainers should come mainly from the local population.

## PROJECT EXECUTION

The *first* phase of the project, which is already under way, includes: (a) the specification of the model framework; (b) the identification and, if needed, modification of training materials; (c) the identification of interested institutions in developing regions with the capacity to support the project; and, in consultation with these parties, (d) the selection of real commodity/natural resource situations for pilot application of the model framework.

The aim is to include five different countries in order to provide a variety of experience for pilot applications. The situations should present well defined problems having a clearly defined geographical scope. They should also share the characteristic of being difficult to solve through existing mechanisms. The project will rely on cooperative institutions in the countries concerned for on-site support, including most importantly, selection of an appropriate situation, liaison with national, regional and local authorities and other interested parties, and logistic support. Suitable cooperating institutions in developing countries are likely to be non-governmental, for instance research institutes.

In the *second* phase, the model framework will be elaborated and translated into concrete blueprints which will allow detailed study and management of the commodity/resource situation selected in each region. This will involve articulating the components of the model framework to correspond with the specifics of each situation and, to that end, will require the close collaboration of the organizations which are providing expertise, including the national/regional institutions. The blueprints will be revised over the life of the project as experience from their application to the actual resource situations is accumulated. The resulting modified blueprints will serve as examples for the application of the model framework to situations elsewhere.

Concurrently with the elaboration of the framework, training will be carried out, involving representatives of the local community, government officials and other stakeholders.

The second phase is expected to last one to three years in each country. Following the conclusion of work in each country, the national/regional institutions will be responsible for follow-up, including continued monitoring of the resource situation and economic and social development using the blueprint.

## PILOT CASE IN NAMAQUALAND, SOUTH AFRICA

Under the present initiative, a first pilot project has been initiated in South Africa, where UNCTAD is cooperating with the Government of the Northern Cape Province in exploring development alternatives for the Namaqualand region.

Namaqualand is located in the northwestern corner of South Africa. It has an area of 47,700 square kilometres and a population of about 60,000 people. The region is semi-desert and has limited potential for agriculture, except for irrigation-based crops on the land along the Orange River, the region's only real river. Goats and sheep herding has traditionally been the basis of the local economy, and a large portion of the population still depends on this activity. However, the number of animals considerably exceeds the sustainable carrying capacity of the very arid land. The main economic activity of the region is mining, with two large diamond mines and one copper mine accounting for most of the formal employment. There is also some tourism, limited, however, to a short period each year when the desert flowers are in bloom. Infrastructure for more extended tourism is largely lacking. Manufacturing is negligible.

The two major diamond mines, Alexcor (owned by the State) and Kleinzee (owned by De Beers) have an expected remaining life of about ten years, after which the deposits will be depleted. As the diamond mines are currently the major employers in the territory, the economic effect of down scaling on the local communities is potentially devastating, and it is thought that, taking into account multiplier effects on employment, the closure of the mines could deprive some 40,000 people, more than half the population of Namaqualand, of their livelihood.

The inevitable and progressive retrenchment of employees from the mines is as much an environmental issue as it is an economic one: experience from previous retrenchment programmes has shown that a forced return to a subsistence existence primarily based on herding of goats and sheep results in immense pressure on the environment. Mass retrenchments from the mines could thus result in increased desertification of an already stressed environment and adversely affect the agricultural potential of the region, as well as its tourist industry.

Economic plight will inevitably lead to the forced migration of currently stable communities that have existed in their villages since well before British colonisation in 1847, and have developed a distinct and unique Nama aboriginal culture and society. The closure of the mines thus materially threatens both the economy, the environment and the social and cultural fabric of the region.

Timely, strategic planning is therefore critical in order to manage the economic, environmental and social consequences of the closure of the mines. At present, economic planning by the authorities and various levels of government is constrained by the resources available to them, and these are under severe strain. Ultimately, decisions regarding the distribution and use of natural resources will have to be made by those whom it affects most — the people of the territory. However, in order for the people to make informed decisions about their future, they have to be in a position to examine options and resolve the conflicts that may arise around the options.

The UNCTAD pilot project is designed to provide the local communities in Namaqualand, as well as other stakeholders, with the tools that will help them to build knowledge and to formulate and evaluate options for land and natural resource use, so that the consequences of decisions can be anticipated and examined. In order to provide for a balanced approach, decisions have to be based on a framework that takes into consideration alternative uses of land and other resources, and assesses their merits on the basis of common criteria such as employment, wealth creation, and conservation. This implies a process of consultation and cooperation involving all who have an interest in the issues at hand, in particular those who have a direct vested interest, such as the project investors and developers, the employees and the communities directly affect by a project. However, in order for all stakeholders, in particular the local communities, to participate on a level where they can develop informed opinions and make qualified decisions, appropriate tools and training are needed. Such tools and training should enable the communities, as well as other interested parties, to examine the various scenarios, and compare outcomes in terms meaningful to them.

The project will be developed with the active participation of the communities through their development fora, and also with other concerned parties such as industry and non-government organisations in the region. It will form one of the elements of the economic development strategy presently under preparation for the province as a whole.

Providing the communities with the relevant tools can be achieved by developing models that represent the linkages between socio-economic activities and various forms of land use. These models incorporate demographic, economic, social, environmental, cultural, political and legal variables, functions and relationships, linked in such a way that developments in one sector will be reflected in effects on other sectors. While the underlying techniques and algorithms are of necessity complex, it is intended that the methodology as a whole be used by non-specialist, trained lay people. The methodology must therefore be interactive, easy to use, and accessible to people in the communities themselves. It needs to be developed using established theory as well as local knowledge, i.e. with the active participation of the affected communities, industries, and other interested parties rather than by specialists in isolation of these parties. It draws on available information, and primary data are collected where there is a lack of key information.

The structural models describe the linkages between production facilities (mines, farms) and the environment and economic structures at the local, regional and national level. The models are based on techniques such as input/output analysis, dynamic simulation, financial simulation (including the flows of income between and within the sources of production, the state and the broader community) and feasibility modelling. They will be successively refined through an iterative process where the interested parties in Namaqualand are consulted and asked to provide their views on both the accuracy of the data and the realism of the functional relationships depicted by the models, thus helping in the task of validating the models. The models are integrated into an overall simulation framework provided by the software platform. Risk assessment techniques are used to analyse the results of the simulations, providing a tangible way of studying policy scenarios and assisting the user in evaluating the effectiveness of development policies.

The data contained in the structural models and the results generated by the functional relationships are linked to a geomanagement system. This system provides an "electronic map", or geographically referenced data management system of the area, which contains the location of natural resources, communities and infrastructure. Information relating to these features, generated by the structural models just described, are stored and linked to specific locations. The tool allows for multiple views or "windows" and simultaneous communication between different sources of information. This provides for easy illustration of results, since the stakeholders can explore alternative scenarios for the future and "see" how these scenarios affect the economy, the environment and ultimately themselves. The softwares used, which are PC-based, are characterized by their transparency as well as by their capability to integrate information in different domains. These features are considered essential in promoting dialogue and they also help to demystify the decision-making process.

Several development options have been discussed in the development fora of the region. These include increased tourism, which would have to be carefully developed, however, because of the present lack of infrastructure and the need to retain the undisturbed landscape. Aquaculture, mainly of shellfish, is another option, as is intensified irrigation-based agriculture along the Orange River. The issue of land tenure is crucial to the prospects of success in this regard. Finally, small-scale mining, which is now carried out at a modest scale, and which was not allowed to develop its full potential under the apartheid regime, might provide significant employment, since there are small deposits of diamond-bearing gravel and, more importantly, extensive potential for semi-precious stones in the region. Scenarios covering all these options will be generated and studied using scenarios.

Apart from development of the models, training is a fundamental component of the project. Capacity building is achieved through the effective use of the tools not only by the communities themselves but also by public and private sector officials and other interested groups. A major objective is to facilitate a cooperative approach to development through the use of a common model at different levels of planning. A further objective is to empower communities and other parties to become closely involved in an integrated decision making process. The capacity building component of the project is carried out through training programmes designed to train trainers. A selected group from collaborating agencies and organizations will participate in the training. These trainers will develop the capacity of the organizations and groups which they represent to successively reformulate and monitor the implementation of development policies.

The training will consist of two parts. The first part focuses on the conceptual framework of the project and will provide a basis for understanding concepts such as sustainable development and how modelling techniques can assist in policy formulation. In the second part, the participants will be introduced to the tools, that is, the geomanagement and simulation software. The training will be based on practical examples, giving participants an opportunity to acquire hands-on experience in exploring different development policies. Policy evaluation and the implications for long-term management of natural resources will be an important component of the training.

## FURTHER APPLICATIONS OF THE FRAMEWORK

It is hoped that the employment of a participatory framework in Namaqualand will help to promote its use in other developing countries faced with the problem of ensuring that mineral projects are developed, implemented and brought to a conclusion in a manner that is consistent with the interests and aspirations of all the parties concerned. In many cases where interests are seen as being diametrically opposed, as is often the case in areas affected by intensive small-scale mining, the adoption of a common frame of reference and the joint exploration of options may result in the identification of not just mutually acceptable, but mutually advantageous solutions.

The UNCTAD secretariat is presently looking for other sites and situations where the framework could be usefully applied.

# POLICIES FOR ARTISANAL AND SMALL SCALE MINING IN THE DEVELOPING WORLD — A REVIEW OF THE LAST THIRTY YEARS

*John Hollaway*
PO Box 5438, Harare Zimbabwe

## INTRODUCTION

The fact that two phrases, not one, are needed to define the subject of this paper underlines the confusion that surrounds all questions of artisanal mining and small-scale mining. In the 1980s a string of conferences largely sponsored by the United Nations spent a disproportionate amount of time trying to define what it was they were talking about. At what point does 'artisanal mining' become 'small scale mining'? When does a small-scale mine become a 'medium scale mine'? The problem persisted across languages: in French the difficulty was defining between '*les opérations artisanale, semi-industrielle et industrielle*'.

No one criterion is sufficient; for instance tonnages depend on grades and market prices. A gemstone mine treating one ton a day, a gold mine treating ten tons a day and a coal mine producing a hundred tons a day can all be artisanal/small scale mines. Such mines may employ, or attract, a handful or tens of thousand of workers, so the size of the workforce is also an unsatisfactory criterion. The value of production may be enormous, as in Brazil's Serra Pelada which generated over $200 million in one year at its peak, or tiny, as with an isolated African woman panning for a few specks of gold on a remote stream. The level of mechanisation may be negligible, as with nearly all gemstone operations, or quite sophisticated, as in Tanzania with its widespread use of compressors, jackhammers and home-made ball mills. This has meant that countries attempting to define small scale mining arrive at having to incorporate specifics in their laws on such matters as the use of wheelbarrows.[1]

Yet, legal reasons aside, there is a need for definition. It will be necessary to return to this issue later, but for the purposes of this paper

— and perhaps for wider adoption — artisanal mining is understood to refer to illegal mining, that is the winning of minerals by people who do not have the sanction of the authorities to do so. This does not necessarily mean that such activity is unsophisticated, although typically it is practised by people with very limited resources. There are technically illegal small mining operations underway in, for instance, the Philippines, with a full complement of shafts, hoists, mills and so on.

Small scale mining, on the other hand, is legal, the miners do have the right to the minerals they are selling and their activities are known to and overseen by the government in the form of mines inspectors and the local authorities. Yet these mines need not be sophisticated; there are numerous formal claims in Zimbabwe which are being worked with nothing more than a pick, shovel and washing pan.

Despite the semantic confusion of the past twenty years, a pattern of policy evolution can be discerned. This development has been hindered by a number of setbacks and false trails, but the position that has been arrived at is remarkable for its widespread acceptance; there are relatively few countries in the world nowadays whose response to an artisanal discovery in a new area is to send the police or army in to chase them off.

However, no country is the same, and with about 150 developing countries in the world it has been necessary in this paper to give a very broad brush picture of the evolution of policy on artisanal and small-scale mining, based on perceived rather than documented trends. Thus it should be understood that there will be exceptions to every statement this paper contains.

## THE EARLY YEARS

All mining was, of course, once true artisanal mining, becoming (in the context of the definition adopted for this paper) small-scale mining as the need to formalise the ownership of minerals developed. While salt, clay and other construction minerals were, and are, commonly mined in a small-scale fashion, the main attraction as always was gold and gemstones. The gold of Africa and the silver, gold and platinum of South America, amongst others, attracted huge populations into the areas where discoveries were made.

However, without explosives or crushers, and in particular without pumps and ventilation systems, the deep mining of hard rock was very difficult, and the miners sooner or later returned to agricultural or pastoral activities, less a hard core who washed for gemstones or searched for promising outcrops to develop. In any event these local miners blazed the way for entrepreneurs from the developed countries, who commonly found no lack of diggings to guide them to deeper orebodies.

As technology advanced and mines became ever larger, more mechanised, more sophisticated and more expensive, the small-scale miner was sidelined. From being a legitimate, if minor, player in the rural economy, he began to be seen as an obstruction to investment, occupying ground that might be better developed by a major mining company, working dangerously, unproductively and environmentally destructive.

All this may not have been important until the 1960s, when the leadership of a very large number of countries, many of them newly independent, were attracted to the socialist-communist-authoritarian system of government. Because the adoption of these policies cut them off from sources of international capital, and because governments wished to control all economic activity, it was necessary to construct siege economies, with price controls and a ring fence of currrency controls to prevent capital flight.

## THE YEARS OF REPRESSION, MID-SIXTIES TO MID-SEVENTIES

In this period many developing country governments, few of whose members had any contact with mining, were intrinsically hostile to the mining sector of whatever scale and whoever the miners. A key phrase in this period was 'the need to re-establish sovereignty over our resources', which was interpreted by many as a code for the nationalisation of existing mines and the creation of state mining corporations to run them. In addition, much misleading technical advice was obtained from socialist countries who had made heavy industry the heart of their own development programmes. As a result attention to be focused on the potential for what would be now called 'world class' deposits, and local small scale miners were further marginalised.

Although initially the free schooling, medical attention and other services provided by authoritarian governments made them relatively popular, along with, in many cases, the euphoria of having thrown off the colonial yoke, the profound unwisdom of poor, undeveloped countries trying to run every aspect of the economy soon began to catch up with reality. Despite considerable development assistance there was soon an intensification of rural poverty, often exacerbated by rapid population growth.

At the same time, the withdrawal from the international financial system resulted inevitably in vigorous black markets in hard currency. The parallel economies that developed meant that those lucky enough to have access to foreign exchange could afford the luxuries (and sometimes necessities) that were unavailable to the rest of the population. Gold and gemstones are, in effect, hard currency hidden in the ground, spurring 'barefoot prospectors' to seek them, and soon, in many countries, artisanal — i.e. illegal — mining had come to be a significant rural activity.

The government reaction to this was usually unequivocal; the state owned all mineral rights, so artisanal miners were stealing from the state. The state had control of all foreign currency, so artisanal miners were sabotaging the economy. Any response short of driving the miners from their diggings and arresting as many as possible would be weak-willed temporising. Direct force was therefore employed (or threatened) in countries as diverse as Indonesia, Tanzania, Mali and Brazil.

However, once the security forces had left the miners would invariably return, and output soon recovered to previous levels. Such were the rewards of artisanal mining that it was usually found necessary to permanently station police or army personnel in the mining areas. A not uncommon result was that the security forces themselves became involved in artisanal mining, usually striking deals with the miners on a tribute basis. Again, after a dip in output when the forces first moved in, production soared, while, in addition to the police and/or army, an everwidening circle of local officials became beneficiaries in return for turning a blind eye. Rumours circulated of very high persons being involved — at cabinet level, even — and the question of how to suppress the artisanal mining sector became a difficult and tainted aspect of government policy.

## THE YEARS OF SEGREGATION; MID-SEVENTIES TO MID-EIGHTIES

Faced with this quandary on artisanal mining, numerous governments asked for assistance from the donor community. It was hardly an area in which most donors had direct experience, but as it appeared to be at bottom a legalistic problem, and as they had ample resources of lawyers, a large number of legal studies were made, often in conjunction with an overhaul of the mining law inherited from colonial times.

The conclusion of this work was that artisanal mining and large scale mining were incompatible. The received wisdom became that artisanal mining was as 'a process set apart from the 20th century concept of mining'.[2] As such it had to be guided by Government. The reference given describes the Tanzanian Mining Act of 1979 as a 'good example of a modern mining law which tackled the problem of small scale mining'. In this law the Minister of Minerals is empowered to designate areas for small scale mining only. However, if a small scale miner uncovers a large deposit then the mechanism provided by the Act allows for such claims to be cancelled by the Minister through his revoking the 'designated area.'

This type of law not only conferred immense discretionary powers to the Minister and undermined the value of a mineral discovery for the local people, it also assumed that (generally poorly paid and unskilled) government geologists were capable of selecting 'suitable areas for small scale miners'. The impossibility of doing this has been recently demostrated in three different countries in Africa, where the writer has had a direct knowledge of the situation. In Kenya in the 1980s a survey of the Turkana region in the north of the country by British geologists led them to state that there was no evidence of gold in the area, yet in 1993 the local tribes people began to discover extensive gold deposits, alluvial, eluvial and primary, such that their output (by dry winnowing) was of the order of 4 kilograms a week. In Mozambique in 1992, the Maniamba cordilheira near Lake Malawi was found to be a major source of alluvial gold and the ensuing artisanal rush led to the finding of significant reef deposits. In Tanzania in 1994 a major new gemstone region was discovered by local people at Tunduru, in the south, of a scale and quality that threatened the stability of the international sapphire and ruby gemstone markets.

## TOWARDS INTEGRATION—MID-EIGHTIES TO THE PRESENT DAY

The failure of the segregated mining regime laws left the challenge of workable strategies for artisanal and small scale mines unresolved, at a time when the Brazilian gold discoveries were producing indelible scenes of human walls of mud-coated miners struggling up the sides of the Serra Pelada pit with their sacks of gold ore.

In addition, concern over the environmental effects of small scale mining was mounting, in particular about the long term effects of mercury. Pollution was reported from a world-wide range of countries[3] leading to further pressure to resolve the situation.

Fortunately, the trend towards financial liberalisation, at first supported and then enforced by the IMF and the World Bank in a large number of countries with an artisanal mining problem, led to the involvement of persons familiar with the mining industry as a business, which is, in the first instance, one of mineral property rights.

The reason for this is that the discoverer of a mine is almost never the eventual owner of it; with only two exceptions there are always transfers of ownership of the title, often several, until it ends up in the possession of a company or organisation willing and able to develop it.

The first exception is where the geologists of a mining company are granted a concession area and make a genuine new discovery of a size of interest to them. This is rare but there have been cases where this has happened; De Beer's diamond discoveries in Botswana are an example. However, it should be noted that this was preceded by fifteen years of exploration on very extensive concessions.

The second is at the other extreme, where the artisanal miner makes a find under a regime where he is forbidden to acquire transferable title — as with most of the segregated mining laws. Thus he has no option but to develop it himself, with resources and skills that are completely inadequate for the task.

Once all mining falls under the same set of simple, transparent laws for the acquisition of mineral title (typically by pegging and claim), a local prospector will usually precede the small-scale miner(s). Prospecting is often an activity undertaken by subsistence farmers during a lull in agricultural work; towards the end of the dry season, for instance. He or she almost certainly has no formal training, no mining skills in the conventional sense and no access to capital. If such a person has a long-term commitment at all to mining, it is as a prospector, in the hope that he or she will be able to find a deposit that others, with more need and/or skill and/or capital will work on for a fee.

This important distinction between the ambitions of the typical seeker of a deposit and those who ultimately work it applies to mines of every size. For this reason the need for an effective legal transferable property right over a discovery by such a person is therefore very important, and not just for 'formal' sector exploration. This right should enable the title to the deposit to be sold, tributed or optioned in a way that is routinely administered, and requires little discretionary action on the part of the authorities.

This assurance of a property right is the driving force behind legal prospecting by local people for economic deposits, and such a right must be sustainable against the pressure from larger companies wishing to acquire the discovery. A primary requirement for a successful small-scale mining sector is therefore a simple, uniform mining code that applies to every scale of potential discovery. Since it is local people that find the majority of deposits of many economic minerals, it might be said that this requirement is a sine qua non for a successful mining sector in a developing country.

By the early 1990s a consensus was emerging that a policy of treating artisanal miners as a marginal minor sector that could be, in effect, legislated away had failed. A major policy change was signalled at a conference sponsored by the United Nations on 'Small and Medium Scale Mines' in Zimbabwe in 1993. The 'Harare Guidelines'[4] endorsed by the delegates and issued at the end of this conference noted that it was not necessary to split up the sector, and hence there was no need to define the scale of such mines; all mining should be undertaken under one legal system with common, transferable rights to mining titles.

In 1995, the World Bank also sent a clear signal that the outlook is changing. It hosted a 'Round Table on Artisanal Mining' in Washington, and invited about 60 representatives of international agencies, NGOs, Government institutions, small scale mining groups and the private sector with interests or concerns in this field. In the keynote speech of Dr. Richard Noetstaller, who authored the World Bank technical paper No. 75, 'Small-scale mining: a review of the issues', he set a clear objective — the transformation of artisanal mining into formal mining, of whatever scale.

Such a policy implicitly acknowledges that there can be no clear boundary between the mines of different sizes, and a healthy mining industry in any country will have a continuum of all sizes of mines, all working under the same legal and fiscal regime.

This theme was echoed by Mr. Peter van der Veen,, the Chief of the Bank's Industry and Mining Division. The problem from the Bank's viewpoint was no longer how to define the sector, for there was no intention to marginalise it; indeed the challenge was how to bring it into the mainstream.

One response to this argument in the past has been that by granting title to local discoverers of mineral deposits, big formal mining companies (by implication big foreign ones) who would like to develop them can be held to ransom. Yet at the meeting evidence was led from several countries, including Brazil, Venezuela and Ghana, that it was easier to deal with people who had formal mining title than with informal groups of artisanal miners made nervous by their lack of legal standing, and by the prospect of having their livelihoods removed without any reward.

The outcome of this meeting was a document which detailed the formalisation process.[5] Coming first, in the policy incentives seen as necessary to achieve this, were the following: the right of access to prospective ground, the right to legal title and security of tenure, the freedom to sell mining rights and the freedom to market the products of the mine.

## CONCLUSIONS AND THE CHALLENGES OF THE FUTURE

This paper has traced the evolution of policy towards artisanal and small-scale mining over the past three decades, from a time when there was no need to have a strategy because the sector was so small, through its starting growth in the 1970s and eighties, and the initial responses of repression and segregation. Neither policy was successful and it has slowly become understood that legitimisation and integration with the mining community as a whole is the best way forward.

This may not appear an easy task, because most artisanal miners have only short-term objectives; these are in essence to earn enough today to be able to live through tomorrow. Small-scale miners, in terms of the definition used here, are a little better; they at least have a secure mineral little (usually for a known period) and hence an asset that can be traded.

There is, however, a further reason for small-scale miners being more successful; they not only understand the need to hold the legal right to a discovery but have had the sophistication and skills necessary to acquire it. There is some evidence that most artisanal miners are not interested in, or consider themselves lacking the capacity for, acquiring formal status. For instance, in the case of both Merelani tanzanite in Tanzania and alluvial gold in Zimbabwe, areas that were being worked by artisanal miners were made specifically available to them for pegging. To the surprise of the authorities in both countries, only a very few miners came forward to claim this right. This simplifies the integration process, as it turns out that most artisanal miners are usually agreeable to being employed by (or more often, contractors to) small scale miners. This turn replaces the artisanal anarchy with a clear chain of responsibility to the small scale miner for governments to enforce the laws and regulations.

Nonetheless most discoveries are too small to attract the larger mining companies, and for most small-scale miners it will be necessary to generate their own resources for development. Once the legitimisation or artisanal mining has been accepted as the objective — and these, as it has been seen, is widely recognised — question of finance and technology loom. It is in these areas — of innovative ways to raise equity or loans for small mine development and of ensuring technology transfer — that are likely to predominate when governments and donors consider the best ways to support small-scale mining in the coming years.

## REFERENCES

1. See for instance, the mining law of Mali. This consists of the Ordonance No. 91-065 (the Law) and the Decret No. 91-277 (the Regulations) and the industry is split into three different sectors under three different sets of Laws, (Orpaillage, petite-mines and mines), using equipment usage as the criterion.
2. Legislative and economic policies for the promotion and regulation of small scale mining in an age of international mining capital. S.C. McGill. UNDTCD, 1984.
3. See for instance: Report on the workshop on ecologically sustainable gold mining and processing. UNIDO Project No. XP/INT/95/043. Jakarta, November 1995.
4. United Nations interregional seminar on guidelines for the development of small/medium scale mining. United Nations, New York. March 1993.
5. A comprehensive strategy towards artisanal mining. The World Bank, Industry and Mining Division, Industry and Energy Department, Washington, D.C. August 1995.

# GENDER ISSUES IN MINING—EFFECT OF RETRENCHMENT

*Paramita Aich*
Research Worker, National Institute of Small Mines (NISM), India

## INTRODUCTION

Retrenchment in mining always adversely affects and redefines people's entitlement and access to socio-cultural, economic and environmental resources of the area. With it, components of impoverishment set forth which manifest in unemployment, loss of land, insecurity in providing food, marginalisation, resource depletion, environmental degradation, migration, social disarticulation and delinquency. The objective of this paper is to assess the effects of retrenchment among different sections of the affected people and the women in particular, which came to light during a field study carried out in this area in 1994–95 by National Institute of Small Mines on the impact of mining on the local people. A case study involving the Bauxite belt of Lohardaga is presented to illustrate the impact of retrenchment.

## IMPACT OF RETRENCHMENT ON LOCAL PEOPLE

The area under study lies in Lohardaga district of Bihar, covering 28 villages spread over 3 Community Development Blocks of Kisco, Senha and Lohardaga. Physiographically, it lies within Lohardaga group of Hills in north-western part of Chotanagpur Plateau. The area is inhabited predominantly by Oraons and also few Mundas. Some primitive tribes like Asurs and Nagesias also live in some remote corners amidst forests. Exploitation of bauxite ores started here before independence of 1947. The introduction of mining in this backward area initially led to a large-scale displacement of local people from their ancestral villages. At the same time, opening up of mining operations ushered in an alternative avenue for earning, with consequent economic upliftment of the labourers employed. But erosion of people's traditional culture, values and identity could not be checked, along with some adverse impacts of mining on environment.

Presently, two big mining companies are operating here, viz, INDAL and HINDALCO. Besides, there are a number of tiny bauxite mines owned by some local wealthy merchants. But a few years back most of these privately owned small mines and a few of HINDALCO's were closed down on the ground of environmental degradation, as reported by the local people. This has led to a mass scale retrenchment of the labour force, mostly the local tribals. In this background, where mining acted as a catalyst for income generation for the last couple of decades, retrenchment of labourers due to closure of mines has accentuated the already existing problems by multiplier effect. This unwarranted retrenchment has negative impacts on the local economy and on the social fabric of the area as well. But its effects are felt in different ways by different groups. Presently, two distinct groups amongst the retrenched mining population can be marked who may be classified on the basis of land ownership, namely:

1. Small and Marginal Farmers
2. Landless People.

## Small and Marginal Farmers

The *small farmers* are the better off amongst all the retrenched people. The majority of this group worked as "miners" in HINDALCO for not less than 15 years. As permanent staff they enjoyed a handsome salary with all kinds of associated benefits. Many of them have invested their savings on purchase of agricultural lands and a few even acquired urban properties in and around Lohardaga. Interestingly, the Christian miners, who have reaped the benefits of education acquired as the result of long missionary influence, have spent a major part of their Provident Fund (PF) money and savings on children's education. A few miners even used it to obtain jobs for their sons. There is a growing tendency among the educated children of the miners to settle for some non-mining and non-agricultural occupations in the urban areas. Most of the miners however, who were about to retire have taken to farming as they still adore this traditional occupation. But the younger group of retrenched miners who are not adequately educated want to utilise their skill in mining operation elsewhere. The present situation has affected them if not economically at least psychologically.

The majority of the retrenched mining families belong to the category of marginal farmers who worked either as "*Hazri*" labourers (daily labourers) in the small privately owned mines or as "*Mazdoors*" (unskilled labourers) in HINDALCO. Their earning was never big enough to encourage the habit of saving and thus purchasing additional land or any other productive assets. The income was just enough to cause neglect of their existing agricultural land, which in turn became cultivable wasteland. The material culture developed by the tribal male folk, based on cash earning from mining operation may be responsible for eroding their passion for nature and particularly, farming. Now that the mining operation has stopped, they have no other way but to opt for cultivation for mere sustenance. The agricultural return here largely depends on (i) quality of soil (ii) physiographic setting of the village and (iii) access to irrigation facility. So the worst hit are those who own proportionately more "*tanr*" land (upland with inferior quality of soil) with poor irrigation facility. The unsatisfactory return from their agricultural land compels them to take up sale of firewood as a subsidiary occupation.

Those who opted for petty business, now repent for squandering away the money. Some even didn't think twice before mortgaging their land to big farmers to venture into trading small articles in mobile stalls or trinket shops etc, which are hardly profitable propositions in the rural areas. With no land at their disposal, they still have to pursue this occupation, which failed to restore fully their past economic solvency.

## Landless People

The male folk who now represent *landless class* among the retrenched families primarily used to work as "*piece-rated*" labourers in the tiny privately owned mines. Most of them assert that quite often their employers took full advantage of their ignorance of labour wage rates to cheat them. Despite such deprivation they feel that their economic position was better then than the present post-retrenchment situation. Now that these people are left with no productive assets, they are forced to leave their habitat, for alien places in UP, North Bihar, West Bengal and Tripura to work in brick kilns as day labourer at a very low wage rate.

A few from the villages near Lohardaga try to work as casual labour in the crowded construction sector. But here the local Muslims always get a preference due to their traditional expertise with the job. Besides, some also seek work as day labourer in different Government employment generation schemes under "Jowahar Rojgar Yojna" (JRY) like repair of roads, digging of wells or construction of houses of fellow villagers.

The majority of the Oraons and Mundas who are basically agrarian communities also work as agricultural labourers at wage rates varying between Rs. 20 to Rs. 25. But due to surplus labour in agricultural season and scarcity of cash, they follow the practice of mutual of exchange of labour instead of charging daily wages.

Those landless people who prefer to stay back in their ancestral land, turn their eyes towards forest, with which they have lived in sustainable manner from the time immemorial. In spite of their instinctive foresight and appreciation of the forest as an integral part of the eco-system, they under circumstance of the present-day changed socio-economic status, have no other alternative avenues left but to take recourse to exploitation of forest as a living by way of sale of firewood.

Thus the closure of mines, on the ground of environmental degradation, has accelerated environmental destruction in a different way i.e. by massive felling of trees. I hope, there is no need to mention separately the consequences of intensive use of forest resources in a region with a fragile bio diversity. Many will no doubt accuse those poor tribals as culprits for destroying the forest. This is, however, the old game of blaming the victim which already had infiltrated into their traditional societies due to long contact with the non-tribals.

## IMPACT ON WOMEN

### Basic Background

It has always been assumed that "household" or the "family" is the smallest unit of convergent interest, where the benefits and burdens of existing conditions are shared by all its members. But there is evidence to the contrary that points to the fallacy of this assumption. The interests of the poor are always neglected, but the women and the female children particularly are affected more with less benefits and more burden. This happens because of intra-household inequalities in regard to health, nutrition, workload and work-area. And the disparities tend to get aggravated at times of economic stress. Of course, it would be unjust to club together all women into one homogenous category. Their experiences, even as women, would be different depending on their economic status, perception and ethnic identity. However in the family context, as women they suffer greater vulnerability even within their own social groups.

The severe outcome of retrenchment on the working families as a whole, has already been indicated in brief in previous paragraphs. But the effects are more mentally and physically traumatic for women. Since retrenchment is a nightmare for all those undergoing it, the question may be asked how does it affect women differently? Or, why should there be any need to focus on women separately? The answer lies in the fact that with retrenchment separate roles for women, based on gender prejudice have been accentuated in the form of various social inequalities. Such inequalities already had infiltrated into their traditional societies due to prolonged contact with the non-tribals. This explains greater vulnerability, lowering of self-esteem and sometimes even a decline of opportunities and status of women of the better off mining families. With this perspective consequences of retrenchment on women are discussed hereafter.

## Changing Socio-economic Status

The economic position of the retrenched miners, who have been able to acquire some land of late, is not so bad as compared to those, who are either marginal farmers or landless. The women of these families with the higher income always assume a sense of superiority in status in their own eyes and also in the eyes of the community. As expected, the degree and nature of participation of these women in the outdoor activities is very low. It has been found that with escalation along the economic ladder the work areas of these women become more and more restricted within the four walls of their houses.

This is more in keeping up with the custom followed by the women of well-off families of non-tribal society. In this context it may be mentioned that amongst the tribals such condition never existed before. The gender equality amongst the tribals in sharing the work load, both inside the household as well as in the outdoor works, was something of an exemplary nature, unheard of among the so-called advanced societies. Hence, ways of adjustment of the women of these families to the present condition are rather painful and different from the women of same ethnic group but of different economic status.

Prior to retrenchment, the role of the women belonging to the families of miners in agricultural sector had been more like supervisors — monitoring the work of agricultural labourers, work assistants, maintaining record of the farm produce etc. But now, many of them are found helping their husbands in the actual agricultural operations because the present monetary situation doesn't always allow them to engage so many agricultural labourers and domestic helps as before. This adds extra burden to their already existing work load at home. The adolescent girls now help their mother in cooking, to reduce the latter's burden of work. In this changed economic circumstance with a drastic fall in the family income, these women are afflicted with more psychological problems. Their sense of propriety gets hurt if economic stringency would have enforced them to depend on forest resources so much that sale of firewood in the open market is perceived as demanding. None of these women have been found to accept jobs outside except their educated sons and in very few cases girls, only if they are Christians. They seem to be quite satisfied with their subordinate status as women. But women approaching middle age show better psychological adjustment to post-retrenchment critical situation, if their children are well placed is secured jobs like in army, public sector enterprises etc.

## Attachment to Agriculture

In the post-retrenchment phase, majority of the wives of the "*mazdoors*" and "*Hazri*" labourers have taken up farming, along with their husbands as the primary occupation. It is true, many of these women, barring a few, didn't cultivate their land wholeheartedly, when their husbands worked in the mines. Surprisingly, the earning from mining was just enough to maintain themselves precariously above the misjudged poverty line. From our perception, we may treat them as lazy and lethargic, but it is their inherent attitude towards life which enables them to be relaxed when they have enough for the moment to satisfy their immediate needs.

Presently, many of these women are facing a new problem. They are being gradually marginalised from their traditional work area by the male folk. It may be due to the newly developed outlook and habit, which their male folk have imbibed through long exposure to outside world due to mining. Majority of these men folk now think that their age-old ecologically sound agricultural system is primitive. So they are fast replacing the indigenous paddy seeds and locally available bio-fertilizers with HYV (High Yielding Variety) seeds to boost up production. Since ages, tribal women were considered as equal work mates of their spouses in agricultural matters where their opinion, experience and expertise in sowing, nurturing and harvesting staple and other crops were taken into consideration seriously. Now with the introduction of new agricultural inputs, which have yet to be assimilated, these women find their age-old expertise futile. In cultivating HYV seeds and using other inputs they follow blindly whatever is told. Thus, these unfamiliar agricultural techniques have dispensed with traditional skill and experience in handling local agricultural seeds and transforming them into mere labourers in their own lands. Likewise, HYVs are also transforming the male folk into mere consumers of purchased seeds and thus increasing the cost of production. Unlike men, women still cognise land as a valuable asset, handed down to them by their ancestors. Hence they think it is morally binding to preserve it and hand it over to their descendants. Of late, in lieu of growing HYV paddy, fervent efforts are being made for cultivation of indigenous varieties of maize and leguminous pulses (*arhar, urad*) hardy mill (*gongoi, madua*) oilseeds (*mustard, sarguja*) in *tanr* lands and winter vegetables in backyards of their houses to maximise food security and ensure crop rotation. This reveals rationality and sensitivity of the women. Here it will be apt to quote "Without the land, the people are lost, without the people the land is lost."

## Influence of Forest

The life of the tribals is geared around forest. They procure everything they need from forest such as food, fuel, housing materials, herbal medicines to mention a few. Even their social, cultural and religious identity are also interlinked with forest. They show a reverence for the forest by worshiping trees like Sal, Karam, Mahua, Mango etc. To check over exploitation of the forest, they evolved taboos and prohibitions concerning the use of forest resources. Rules were also formulated for proper use of individual species of trees and other forest produces like the trees of high economic values could not be felled under any circumstances except on the special ritual occasions. In this way, a symbolic relationship between the tribals and the forest continued for long. But the situation has changed today considerably. The collection of fuelwood and minor forest produces, had always been the duty of the female folk for domestic consumption. But the present post-retrenchment impoverishment and monetary scarcity have compelled the female folk of the families of "Hazri"/"Mazdoor" labourers and "Piece-rated" workers to collect more firewood than before for sale, to complement their meagre family income. In many cases, these women are even found to help their male folk to fell trees of preservation status and restrictive use, to augment the quantum of firewood to be sold in the market. It is true, that many of these women feel psychologically disturbed for rapaciously exploiting the forest, but they have no other way but to adjust to this present desperate situation.

Presently, a woman of these retrenched mining families on an average devotes 3 days a week for collection of firewood and 2 days for selling it in Lohardaga market. Prolonged deforestation both by outsiders and to some extent by local tribals is responsible for recession of forest line from the immediate vicinity of the villages towards the interior. Consequently, the distance covered to collect firewood has increased from an average of 1.5–7 kms. Similarly, they have to travel about 8–9 kms to reach Lohardaga to sell head-loads of firewood at a very low price of Rs.15 to Rs.17. Thus the number of human days spent, work load and hardship involved in collection and sale of firewood have more than doubled now than before.

Though male folk work along with women in collection and sale of firewood and in cultivating their own small farms they seldom show any urge to help the women in collection of water, washing, cleaning and cooking for the family members. So these have always remained a woman's responsibility exclusively. After a back breaking long hard day's work, both domestic and out door for extra income, how can a woman have time to devote for herself either for rest or leisure?

## Brick Field Worker

Many women belonging both to the families of landless people and the marginal farmer family accompany their spouses to the brick-kilns to work as wage labourers. The brick-kilns are the most exploitative sources of livelihood in the unorganised sector that offer no protection to the workers. The local "Sardars" quite often working as labour agents persuade and help them to migrate to the brick-kilns. There they, along with their husbands, live under abject poverty and wretched conditions. The women in this profession are more often underpaid and ill treated. They work at a very low rate, lower than the rate prevalent for the male which itself is always less than 50% of the stipulated minimum fixed by the government. However, earning from such hard labour is barely sufficient for their families to tide over the winter season, when agricultural work in the native village comes almost to a halt, but they cannot save adequately to bring back home.

The common worries voiced by the women to the study team are regarding chances of molestation and eve teasing by local people and contractors at brick-kilns who take full advantage of their simplicity, lack of inhibitions and poverty. All this with added violence, affect their physical and mental health and is accompanied by a high risk of contracting AIDS.

On the other hand, such seasonal migrations to brick-kilns located near urban centres have given them greater freedom. Such freedom in the new surrounding has induced greater economic independence and confidence in their tone. These newly developed qualities are beneficial, particularly for those women who prior to retrenchment were confined within their habitat. Thus, it has helped them to regain the old status enjoyed by tribal women traditionally.

## Bias against Young Girls

A few of the poor women of landless families in a complaining tone, have reported to the study team, that always a boy is preferred for most of the periodic work schemes of the government, under JRY, which is quite ignominious for the section of the tribal women who still cherish the thought that their society is egalitarian.

The children, especially the girls of retrenched mining families are in a way the worst affected as their future suffered a set-back. The rate of school dropouts among girls has increased suddenly with retrenchment. In the opinion of the teachers interviewed, the reason for the poor enrolment of girls lies in the fact that they are to look after household chores, as their mothers are to undertake the burden of extra work outside. Ever and anon, they are also found to accompany their mothers to forests for collecting dry leaves, or herding cattle of others. Going to schools is a luxury for them, as the time spent in school can be well utilised to give relief to their overstrained parents by lending helping hands. This biased attitude of people towards girls reveals gradual marginalisation of women based on gender inequality.

## Health Status

Given high mortality rates among women, it is likely that they will be the worst affected by retrenchment induced morbidity. Similarly, the nutritional and health status of the woman, which is lower than that of the males even under normal circumstances, is bound to proportionately go down in the event of an overall decrease in health status caused by retrenchment. The per capita intake of calories will surely show a significant drop amongst women, which will make them more anaemic and undernourished. Thus with poor immunity the risk of being prone to all auxiliary diseases will increase.

## CONCLUSION

The orthodox environment activists may appreciate the decision of the closure of the local small mines for inducing environmental degradation. But, it will not be wrong to say that the authority implementing these restrictions has followed a reductionist approach rather than a much needed holistic thinking. It is this short-sighted outlook that instead of preventing a disease has created a few more. Before adopting any action plan, both the short and long term, consequences should have been taken into consideration. Has closure of mines in the study area checked environmental degeneration? Have unchecked soil erosion, depletion of forest and land degradation stopped with the closure of mines? Our experience say no. What about the human beings, especially women and children? Are they not an integral part of the eco-system? "Marginalisation of women and destruction of bio-diversity go hand in hand." Is it necessary to destroy peoples way of life in the name of environmental protection without offering them any better substitute? Although retrenchment may in some cases be necessary from techno-economic consideration *Rehabilitation Action Plan* is all the more necessary before large scale retrenchment, as it is necessary before starting a mining project or abandonment of a mine.

Impoverishment thus is the common denominator of all the aspects of post-retrenchment phase. The only way to avoid such a critical situation is through sustainable development by learning the nature's law and to master the art of applying them correctly. Constant effort for developing awareness through non-formal education, revival of submerged indigenous ideas and whole hearted participation of the local tribal population at all decision making stages, will help to accomplish regeneration of the environment. But at the same time, one should keep in mind that the principal victims of retrenchment have been the women and children, many of whom are now living on the knife-edge of survival. Policy makers need to understand that continued neglect of women, the primary house-hold managers and care-givers for the society can rebound on an area's economic and social development for decades to come. Henceforth, gender-sensitive policies should be implemented to eradicate marginalisaton of women and revitalise the moribund economy. In this respect I would like to quote what Leopold Kohr wrote in the "The Breakdown of Nations."

"The fascinating secret of a well-functioning social organism seems thus to lie not in its overall unity but in its structure, maintained in health by the life-preserving mechanism of division operating through myriads of cell-splittings and rejuvenations taking place under the smooth skin of an apparently unchanging body. Wherever, because of age or bad design, this rejuvenating process of subdivision gives way to the calcifying process of cell unification, the cells, now growing beyond their divinely allotted limits, begin, as in cancer, to develop those hostile, arrogant, great power complexes which cannot be brought to an end until the infected organism is either devoured, or a forceful operation succeeds in restoring the small-cell pattern."

### REFERENCES

Dhar, B.B. and Saxena, N.C. Socio-economic impacts of environment. New Delhi, Asish Publishing House,1994.

Eckholm, Erik P. Down to earth-environment and human needs. New Delhi, Affiliated East-West Press, 1982.

Sen, Geeti, ed. Indigenous vision : people of India : attitudes to the environment. New Delhi, Sage Publications, 1992.

Fernandes, Walter, Menon, Geeta and Viegas, Philip. Forest environment and tribal economy: deforestation, impoverishment and marginalisation in Orissa. New Delhi, Indian Social Institute, 1988.

# LEGISLATIVE FRAMEWORK IN INDIAN MINERAL SECTOR AND SCOPE FOR ITS IMPROVEMENT VIS-A-VIS SMALL SCALE MINING

*R.L. Bhatia* and S.V. Ali**
*Dy. Controller of Mines, ** Controller of Mines
Indian Bureau of Mines, Nagpur

## INTRODUCTION

The legislative framework applicable to mining industry is quite vast and exhaustive and is subject to a lot of regulations and restrictions covering wide gamut encompassing conservation of minerals, protection of environment, safety, labour welfare, payment of wages, regulation of contract labours in mines, etc. However, there is an ample scope for its circumvention, and enforcement of the legislation is still rather lax. Some of the issues discussed below are the areas which require consideration by the enforcement authorities of the State and Central Government.

## STATEMENT OF PROBLEM

It is desirable that rules should be practical and workable. The law is required to keep pace with the technological advancement taking place.

So far the statutory authorities have adopted a rather conservative and extra safe approach particularly as regards slope angle of benches, height of benches, provision of protective equipment.

Many of these rules are violated not primarily because small mine owners intend to violate them. However, the rules lack rationale. There is a need to bring simplification/standardisation in some of the rules as the use of ANFO where the design of mixing shed could be standardised and can be easily obtained by the entrepreneurs, and it should not take much time in getting the permission from statutory authorities. The Indian labour having agricultural background are not accustomed to wear shoes/helmets, more so, in case of female workers. Many of the workers are also not used to defecate in the enclosed latrines provided in accordance with the rules.

Land acquisition problems are assuming serious dimensions in carrying out mining operations including blasting, transportation, etc. Procedural formalities involved are too many and, therefore, take unduly long time. This may have repercussions on economics of a mining venture. Many a time, labourers do not get due compensation. Small mine owners do not have financial capacity to withstand these delays.

To bring in scientific temper in mining, the government have introduced the concept of mining plan which is a prerequisite for every fresh grant/renewal of lease. While the objective is good, at times the demands made by enforcing authorities are too detailed and many a time not relevant. The small mines are subject to fluctuating demand of mineral, operations are of seasonal nature due to rains, labourers are not available due to sowing/harvesting season, etc. the mine owners experience difficulty in observance of these legal provisions. Many of them are not literate and have neither resources nor will to carry environmental protection measures. A pragmatic approach is called for.

## REGULATING GRANT OF LEASES

Presently, the production capacity available in Indian mining industry in respect of a few minerals is in excess of the demand. As against a number of about 9,000 leases granted, so far only about 3,200 mines are reporting production. In case of minerals such as limestone, iron ore, soapstone, mica, etc., about two-thirds of the leases are lying idle. While small-scale mining is to be encouraged and is to be treated as an aid to exploration subject to certain limitations, there has to be some regulation on grant of more leases in case of minerals where excess mining capacity vis-a-vis demand exists. It is, therefore, desirable that the consideration to the demand that exists of a particular mineral should be given while granting fresh leases except where special conditions so require such as location close to already established mineral-based industry. Otherwise, observance of statutory provisions including implementation of mining plan will be difficult for such new entrepreneurs.

The instances have been seen when a patta has been granted by the state authorities even after grant of lease to certain parties. This not only disturbs the mining operations but these parties also get the bargaining powers for extortion of money from mine owners.

## SHAPE OF LEASE AREAS

Presently, leases are granted according to Khasra numbers. The shape of some of the leases granted is very irregular, so much so, that it is difficult to work systematically and substantial quantity of mineral is not available for exploitation. A regular shape of lease lends itself to systematic mining.

## RATIONALISING ROYALTY RATES

Royalty rates currently in force as per second schedule to MMRD Act, 1957 are required to be rationalised. In case of a number of minerals, the lower limit of the grades have not been specified. In case of iron ore mines, there have been disputes as regards whether royalty is payable on slimes leaving the beneficiation plant and impounded in tailing pond located outside the lease boundary. Similarly, in case of other minerals where crushing and screening plant is located outside the lease boundary, the royalty becomes payable even on the rejects which are not marketable. This also creates difficulty for lessees to dispose their rejects as construction material to local State Government authorities, for such material become uneconomical in case royalty as applicable to major minerals is levied on these rejects. A concessional rate of royalty in such cases will not only free the land in the lease so occupied by these huge reject dumps but will also obviate the necessity of mining of minor minerals from other areas.

## REQUIREMENT OF LEASE AREA FOR INFRASTRUCTURE

As per Sec. 5(2)(a) of MMRD Act, no area is to be granted on lease until and unless there is an evidence to show that the area has been prospected earlier and the existence of the mineral content therein has been established. In case the area is required for infrastructure i.e. crushing and screening plant or for any other purpose by the applicant for mining lease, the mine owner finds it difficult because of the above statutory provision. In case of genuine requirement of the area, a provision is required to be made in MMRD Act for exemption.

## WORKING WITHIN LEASE BOUNDARY

As per the model form of mining leases in Form-K part-7, under MCR, it is provided that the lessee/lessees shall act at his/their own expense to erect and at all times to maintain and keep in repair any boundary mark and pillars according to demarcation to be shown in the plan annexed to the lease. Such marks and pillars are to be made sufficiently clear of the shrubs and other obstructions to allow easy identification. However, in actual practice lease boundary pillars are often found absent and lessees more often than not trespass the lease boundaries without its detection by State Government authorities. A suitable mechanism for proper demarcation of lease boundaries is required to be evolved to check malpractices on the part of the mine owners on this count.

## PROVISION FOR AMALGAMATION OF LEASES

At present, a number of mine owners are having 2, 3 or more leases being operated as one mine. These mine owners are required to maintain separate accounts of royalty to be paid leasewise. Moreover, separate mining plans are required to be submitted for each lease as per the requirement of the Mineral Concession Rules, 1960 (MCR). There is no provision in MCR for amalgamation or merger of leases. The merger or amalgamation will obviate the necessity of maintaining separate accounts and plans/sections for each lease.

## AREA OF LEASES

The area of a number of leases granted is so small that it is very difficult to work systematically. About 12% of leases are of area less than 2 hects. and about 25% have an area less than 5 hects. A minimum area say 10 hects. and above is necessary to carry out work in a systematic way as desired by the statutory authorities.

## RELATIONSHIP WITH OTHER STATUTES

While steps have been taken to simplify and rationalise/liberalise certain statutory provisions after introduction of liberalisation policy of the government since July, 1991, however, there has not been corresponding amendment in other related statutes. For example, as per rule 24(A) (clause 6) of MCR, 1960, the lease shall be deemed to have been extended by a further period of 2 years and/or with the date of receipt of the orders of the State Government thereon, whichever is shorter. In case lease area is under forest, the extension/permission to work is not available in the forest land beyond the lease period.

## RESTORATION, RECLAMATION OF MINED LAND

As per Rule 34 of MCDR, every holder of PL or ML is to undertake restoration, reclamation and rehabilitation of the lease area. There are two aspects which should be taken into account before this rule is enforced. Firstly, continuance of mining beneath existing pit which may not be economical as on date but which may become economical with the passage of time, improvement of technology, etc. Secondly, whether sufficient overburden material is available to fill the void caused due to mining. However, so far except in case of shallow pits of bauxite, gypsum, etc., the rule is not being implemented. As this entails huge cost, consideration to levy some cess may have to be considered which could be utilised for this purpose so that the objective of the rules is fulfilled.

## CLOSURE OF SMALL MINES DURING RAINY SEASON

As per Rule 24 of MCDR, 1988, the lessee has to intimate IBM, when the work in a mine is discontinued for a period exceeding 30 days. Almost all the small mines numbering over 2500 are such that these do not work during rains and also due to seasonal nature of employment. Practical purpose for such requirements needs to be looked at afresh.

## PREPARATION OF E.M.P.

With the emphasis on environment now the lessee is required to give information in environmental impact assessment document not only related to area lying within the lease but also of the area surrounding the lease for which the lessee has no right to access. This area at times may be a reserve forest. This difficulty on the part of the lessee cannot be ignored.

## EMPLOYMENT OF QUALIFIED MINING ENGINEER IN MINES

As per provision of Rule 42 of MCDR, 1988, a lessee is required to employ a part-time or a full-time mining engineer and elaborate duties have also been prescribed in Rule 44. However, in actual practice, it is seen that a mining engineer as required under Rule 42 is not being employed by most of the small lessees since the contribution to be made by him in the mine does not appeal to the mine owners. Naturally, remuneration paid to him will be dictated by such considerations. Moreover, there is a lack of monitoring mechanism on the part of the enforcing authorities and if at all, the rule is being observed merely on the paper rather than in actual practice.

## PRESENT SCENARIO IN INDIAN MINING INDUSTRY

It is, therefore, no wonder that the conditions in Indian mines particularly small mines are still not very satisfactory and they leave much to be desired. A number of mine owners are required to carry out occasional blasting and while explosives such as gunpowder and safety fuse are used, these are not reported to avoid legal provisions. As against about 3,200 working mines only about 1,000 are reporting use of explosives in returns. Still primitive methods like inducing cracks in the strata by burning wood/coal, well like workings are carried out, loading of minerals on ponies/bullocks are being practiced, special gelatin is cut into pieces to economise, blasting operations are carried out by unqualified and uncertified persons.

## CONCLUSION

There is a need to educate the mine owners and workers and the statutory authorities may consider to arrange free dialogue/workshop with them so that legal provisions are appreciated. This will go a long way in better observation of rules particularly by small mine owners.

It is desired that enforcing authorities are adequately strengthened to carry out inspections and pragmatic approach are adopted by them. Rules are made more rationale and simplified so that the better scenario in Indian mining industry emerges.

## ACKNOWLEDGEMENT

The authors are thankful to the Controller General, Indian Bureau of Mines for giving permission to present this paper. The views expressed in the paper are those of the authors and not necessarily of Indian Bureau of Mines.

# CLUSTER-MINING: A TESTED CONCEPT FOR SMALL MINES

S.L. Chakravorty
Hony. Secretary General, National Institute of Small Mines (NISM)
India

**PREAMBLE**

The concept of small-scale cluster-mining as a means of sustained employment and eco-friendly mining operation is gradually attracting the attention of many discerning personalities all over the world. It is a tested concept1. When developed with imagination and long term vision it proves not only to be a big source of productive employment but also convenient in shaping the small-scale mining operations on modern line in eco-friendly manner consistent with safety and protection of environment. The financial investment needed for generating unit employment is also minimal as compared to creation of high-tech industrial employment. Although cluster-mining cannot or need not altogether replace isolated small mines widely dispersed in different parts of a country, it is one of the surest ways of developing sustainable mining operations with high employment potential with a degree of certainty. This concept works better and quicker if the whole operation is technically and administratively preplanned by a Govt. body or a resourceful individual organisation with some authority. But this task needs to be taken up as a mission with responsibility and not merely as a techno-bureaucratic exercise which would merely ensure investment but not result. This effort would in effect generate confidence of small entrepreneurs in commercial ventures and develop a breed of small investors. This would be one of the many practical ways of permanently diverting young educated unemployed from the attractive path of mindless antisocial activities and that too achieved at a low cost.

But the problems of development of such clusters are so diverse and there are so many local problems to be handled (technical and socio-economic) that each country may need indepth study and consideration in each case. A case study of a similar experiment in India (Pachami-Hatgacha) may be found interesting and worthtrying under similar circumstances. In this case an original employment figure of about 200 or so has gone up to over 10,000 in about two decades, providing minimum sustenance to a large number of people. A quiet village centre in the past is now a centre of bustling stone mining activities.

## TWO CATEGORIES OF CLUSTER-MINING

There are two categories of cluster-mining — (i) developed naturally in course of decades of operation and (ii) pre-planned and executed under same authority fairly quickly. By way of example, Pakur Stone mining in Bihar, Makrana Marble mining in Rajasthan and Naini Glass Sand mining in U.P., can be classified under the first category and Pachami-Hatgacha Stone mining in Birbhum in West Bengal and, in a way, Panna Diamond mining under the second category.

Naturally developed cluster-mining takes a long time to develop and gives rise to employment in a slow and halting manner. This happens as and when the number of mines and production gradually increase, sometimes with erratically growing market demand and slow development of infrastructural facilities. In such clusters proper and timely geological investigations and scientific analyses and studies are not carried on in time, leaving the entire operation to hit-and-miss technology except in a few isolated cases of enlightened individuals.

But in the case of preplanned clusters all the necessary preliminary investigations are made and the entrepreneurs helped and guided to carry on operations in an enlightened manner. Thus the development is much quicker on a sure footing leading to fast production and employment. Moreover, in such cases, chances of introduction of more effective mining methods and better technology (Intermediate) is easier and the entrepreneurs are more receptive to suggestions and advice. This was realised during two NISM organised Workshops (1995) on motivating small mine owners for eco-friendly mining operation.

## ESSENTIAL BACKGROUND FOR PLANNING CLUSTER MINING

Since small-scale mining (at least in developing countries) is basically labour intensive and consequently has low productivity, the employment is comparatively much higher for a given output although the wage level is much lower than in the mechanised mines. But fortunately the village level workers used to cheap village level life style, supplemented by agricultural earning, can manage things better with lower wages than those living in unhealthy urban atmosphere in many of the mechanised big mines. Since labour intensive mines do not need much of an investment and involve shallow digging, such mines can be started almost any where if there is adequate demand for the mineral and the mineral occurrence is available within a shallow depth.

But small isolated mines without adequate market, geological investigations and transport facilities are, more often than not, seasonal operations moving with varying and erratic market demand. The position is made worse by the consumers not feeling inclined to patronize because of uncertainty of supply, non-standard materials and often due to bad accessibility. Thus isolated small mines quite often are not sustainable and frequently close down due to inadequate cash flow, discouraging adequate investment.

Against this background of isolated mines what might appear attractive with good chances of success is Cluster-Mining planned with initial technical, administrative and marketing-cum-financial support. But for starting and running such mines on a sustainable basis certain background facilities and factors are essential and important. Some of them are:
- adequate existing market within a reasonable distance from the mineral deposit
- adequate facilities for mineral transport
- availability of some essential infrastructure facilities such as communication, public transport, power line etc.
- away from dense habitation and possible serious environmental problems.

## ADVANTAGES

Some of the important advantages of such clusters are:
- development of common infrastractural facilities at mine site at minimal unit investment e.g. power line and road, railway siding, water supply, telephone, public transport facilities;
- common geo-technical investigation of the entire area at minimum unit cost;
- common technological, administrative, marketing, accounting and advisory support;
- utilisation of costly equipment and machinery on cost sharing basis;
- development of common supply facilities — fuel, explosives, cement, steel, common equipment etc.;
- successful joint resistance to exploitation by middlemen and development of viable marketing strategy so essential for survival;
- common approach to EIA and EMP on cost sharing basis which would be a possible solution to problems of environmental control of isolated small mines, which cumulatively cause considerable damage without any steps for regeneration.

## ENFORCING ECO-FRIENDLY MINING OPERATION

Enforcing eco-friendly mining operation in small mines is one of the most challenging tasks for the administration. Apart from ignorance, high financial cost in carrying out EIA is one of the main reasons for ignoring the environmental responsibilities. As the matter stands today, except for giving full exemption to those small mines having less than 5 hectare of lease area, all the other mines are subject to the same rigours of legislation enforcing same parameters of restrictive provisions as are applicable in the case of big mines. As a result thousands of small mines simply ignore the environmental provisions and do not care about eco-friendly mining operations. The whole situation is a matter of considerable worry and it is not known how to effectively persuade the small miners. Recently, the Govt. of India in the Ministry of Environment and Forest sponsored two Workshops organised by NISM (1995) to motivate small/medium-scale mine owners to carry out mining operation in an eco-friendly manner. The Ministry also sponsored a project study (continuing) by NISM for developing environmental guideline for small mines. The experience of the two Workshops is encouraging. It was appreciated that it would be possible to carry out EIA of a cluster of mines on cost sharing basis and introduce management plan in a realistic manner. It was also realised that motivation is a more effective approach than mere legislation and motivation is much easier in clusters than in large number of isolated mines because in clusters there are always some relatively progressive minds who influence others as catalytic agents. These Workshops thus developed the idea, without any opposition, that it would be easier to enforce eco-friendly mining operation in clusters with full co-operation of the mine owners. Thus cluster-mining is attracting active attention of the Govt. of India also.

## MAKING A BEGINNING

For undertaking such planned development of Cluster Mining certain basic steps are necessary. Chronologically they are:

(i) Identification of a market with adequate demand on sustained basis for any industrial mineral e.g. Stone Aggregate, Limestone, Brick Earth, China Clay or Refractory Clay, etc.
(ii) Identification of the resource area and the authority must keep the area reserved out of reach of speculators.
(iii) Through geo-technical investigations estimation of reserve, depth of occurrence, quality and possibility of upgradation.
(iv) Marking of the entire area into different blocks (at least 30–40 acres each) with adjacent barriers for transport road, stacking ground, office accommodation etc.

(v) Broad determination of the cost of extraction, processing (if necessary) and transportation.
(vi) Working out of a model feasibility report identifying sources of supporting finance and minimum entrepreneur-investment necessary for each unit.
(vii) Working out of an overall plan of the entire area in some details, identifying financial and other responsibilities of the administering authority and how and after what period of time the total investment can be realised.
(viii) Induction of small entrepreneurs, moderately educated and with determination to work as self-employed workers and willing to work hard in the field in partnership or under co-operatives with friends. The choice of partners should be left to the entrepreneurs and not decided by the authority. The authority should, however determine and thus advise about the broad criteria of the entrepreneurs. This is an important advice which may need mild persuasion with the entrepreneurs.
(ix) Offering initial technical guidance for starting mining operation either directly or through relatively more experienced entrepreneurs who can teach others by example.
(x) Offering commercial loan and or other form of techno-financial support for starting operations and acquiring basic minimum equipment. The entrepreneurs themselves should invest, however small it may be, and thus have a stake in the project as otherwise the financial support will be looked upon as a dole which would be disastrous.
(xi) Offering some initial help in marketing by way of advice and support in tendering, arranging supply and realization of sale value. If necessary, initially the planning authority may organise sale after buying up the materials from the entrepreneurs on cash basis. This way not only the confidence of the entrepreneurs would increase, an essential necessity initially, but also their initial cash-flow would be manageable. This arrangement may be continued only for a short period till the entrepreneurs become mature in the tricks of the trade.
(xii) Finally, help in formation of an association of the entrepreneurs for guiding their commercial activities, protecting common interests and meeting common needs. But initial few years may need some control from the planning authority which should ultimately dissociate itself from direct interference in the whole affair and should merely stand by.

## LIMITATIONS OF APPLICATION OF THIS CONCEPT

This concept of cluster-mining can conveniently be introduced for industrial and minor minerals where problems of smuggling do not have to be faced. However, even for high-price low-bulk materials like gold, precious and semi-precious stones the gold-rush-atmosphere could some what be managed or mitigated by the concept of cluster-mining with proper demarcation of the area into manageable small blocks, and defined mining right and monopoly purchase at international rate under reasonable degree of Govt. control. By limiting the depth of such open operation the deposit at greater depth can be exploited by a well organised and well financed mining company maintaining good mutual relationship with the small operators in a planned manner with active support of the Govt. Thus this way the ill effect of illegal artisanal mining and gold-rush-atmosphere can be some what mitigated. In such organised clusters common facilities of extraction of gold or other precious minerals can avoid environmental pollution by better guidance, practice and control.

In India such cluster mining of diamond can be found in Panna in Madhya Pradesh, marble in Rajasthan, Glass sand in Naini in Uttar Pradesh, Stone aggregate in Pachami-Hatgacha in Birbhum in West Bengal. But unfortunately, in most of these areas the total advantage of cluster mining can not be harvested for want of adequate follow up and supportive actions by many of the Govt. authorities who are more interested in collecting royalty income than in improving mining activities consistent with safety and protection of environment. They are not much interested in gradually modernising mining and processing practices which not only improve labour wages but also Govt. revenue income.

## CONCLUSION

This approach in developing cluster-mining for creation of sustained productive employment quick enough is basically an exercise in human psychology and behaviour. And for success what is necessary is clear vision, good planning, and firm handling in a transparent manner. It would be an exercise in motivating, guiding and training in self-employment needing, on the part of entrepreneurs, hard work, hard initial life and learning the basics of commercial venture. Such planning can successfully be undertaken by a Govt. body or a resourceful private organisation with adequate delegated authority under its control, needing only minimal financial resource with no political interference. This type of project attuned to the local conditions can also be undertaken in other developing countries where unemployment situation is acute and NISM is prepared to help.

### REFERENCES

UN Guidelines for the Development of Small/Medium-Scale Mining: Selected Papers; Presented at the UN Interregional Seminar held in Harare, Zimbabwe, 15-19 February, 1993.

The World Bank organised Roundtable Conference on Artisanal Mining, Washington DC, 17-19 May, 1995

# SECTION 2

# Environmental Perspective

# AUSTRALIAN INITIATIVES FOR BEST PRACTICE ENVIRONMENTAL MANAGEMENT IN SMALL SCALE MINING IN THE ASIA-PACIFIC

*Peter Hancock*[1] *and Stewart Needham*[2]
[1] Visiting Fellow, Centre for Resource and Environmental Studies, Australian National University
[2] Supervising Scientist, Federal EPA, Canberra

## INTRODUCTION

Small scale mining is an activity that is better known by its heritage and environmental impacts than by definition of its scale. It encompasses a range of mining operations from very small scale, essentially manual or artisanal mining to small mines of up to 100,000 tpy which may be mechanised and involve a modest level of capital. The Australian federal Environment Protection Authority (EPA) has an interest in addressing environmental management at all levels and types of small scale mining in developed and developing countries in its region. This paper outlines the measures the EPA has taken to foster environmental best practice in mining in general and how it now seeks to work with India and other countries to include small scale mining.

Brazil, India, Indonesia and Zaire are estimated to each have some 500,000 artisanal miners, while in China the number directly engaged in small-scale mining has risen from 5 million in 1986 to 8.85 million in 1994 working some 280,000 mines (Li Peiji and Li Guangwei, 1996). There are now some 11 million artisanal miners globally.

Artisanal and other small scale miners account for a significant proportion of the world's mineral production, including 31% of industrial minerals, 20% of coal, 12% of metals, 20% of gold, almost 40% of diamonds and practically all other gemstones (Noestaller, 1987 and 1995; Labonne, 1996). In some countries this is much greater. In China 43% of the country's

---
[1] Consultant to Federal EPA.

1993 coal production of 1150 million tonnes and most of the industrial minerals were from small scale mines.

In rapidly expanding economies where small scale mining provides a high proportion of minerals needed for domestic consumption and development, governments want to increase production by improving the efficiency of small scale mining. Although productivity and efficient extraction may well be the primary goals, governments and communities now recognise that improved environmental management is required to reduce environmental impacts and rehabilitate mined land to a valuable post-mining land-use. This is an essential component of sustainable economic development.

Many countries now have legislation that requires all mining operations to be licensed and responsible for occupational health and safety (OH&S) and environmental impacts. Legislation alone cannot translate into improved performance in small scale and artisanal mining, except where there is a high level of literacy and environmental awareness. In most situations there will have to be a reaching out to the individual miner with hands-on demonstrations of how they can improve their environmental performance in order to improve their production, profits, OH&S and quality of life. This requires an understanding of the economic, educational, and cultural setting of small scale miners, whether they be cottage industry, artisanal miners using traditional manual methods or miners with powered excavators and recovery plants. In addressing ways to further environmental best practice in small scale mining, the EPA recognises the enormous importance of artisanal mining in terms of its positive attributes:

- gainful employment and relative economic wellbeing by alleviation of poverty for so many
- rapid flow of virtually all gross revenue through wages and profits into the local economy because there is virtually no capital to repay or service
- significant contribution to global mineral output
- extraction of small or low value resource occurrences which would not otherwise be utilised, being uneconomic for medium to large-scale corporate miners.

Similarly, the EPA recognises the negative environmental and socio-economic effects of small-scale and artisanal mining, many of which can be addressed by working towards best practice in environmental management. These include:

- loss of vegetative cover and degradation or destruction of landscapes and ecosystems
- erosion and siltation
- pollution of streams and rivers
- high level of injury and sickness

- poor social, health and safety conditions at mining sites
- inefficient extraction of resources
- gender inequity — almost half of the six million artisanal miners are women, but their share of the benefits is far from commensurate (Labonne, 1996).

Problems and remedial policies pertaining to artisanal mining have frequently been ventilated at international conferences (Burke, 1996; Labonne, 1995; Noetstaller, 1995) and are well known to governments of countries with these problems. This paper looks beyond discussion of the problems and policies to link the issues into solutions with practical measures. It invites developing countries to liaise with the Australian EPA to define country issues. It seeks their opinion on how the EPA can assist with advice and training so that Australia's substantive environmental management experience can be adapted and applied in their country.

Substantial benefits of a more sustainable development of small-scale mining can be enjoyed inn developing countries if environmental and related socio-economic problems and their underlying issues can be addressed with practical systems and measures. It is the aim of Australia's EPA to facilitate this.

## UNDERLYING ISSUES

### Occupational Health and Safety (OH&S) and Community Health

These are priority concerns in developed or developing countries. Best practice environmental management and OH&S go hand-in-hand in mining operations. Many of the negative impacts on the bio-physical environment that arise from poor environmental management also impact negatively on the miners themselves and the local community. The wellbeing of artisanal, and to a lesser extent other small-scale miners, is often compromised by unsafe working conditions. This is commonly caused by lack of knowledge or disregard of the risks. Examples are:

- unstable batters and benches formed in quarries and pits
- unstable underground workings with inadequate support structures causing ground failures
- poor ventilation and lighting, flooding and gas accumulation in underground workings
- damaging levels of dust and noise
- inadequate mining equipment and lack of training
- unstable spoil heaps adjacent to working and living sites
- non-availability of, or failure to use, personal safety equipment
- mercury poisoning of miners using the amalgam process to recover gold and contamination of waterways and food chain with mercury.

## Water Management Issues

Water resources require careful management in any environment to ensure adequate supply and quality for mine and worker use. Good environmental management of water necessitates planning to control the flow, containment and discharge of all mine water and run-off (storm water) to prevent:

- contamination of streams and rivers with suspended solids, increased acidity and heavy metals in solution that may affect aquatic life and downstream communities (e.g. suspended solids, acid generation, heavy metal concentrations, mercury contamination)
- contamination of aquifers
- erosion, siltation and loss of vegetative cover from run-off and the damaging practice of ground and river bed sluicing
- discharge or spillage of oil, reagents and human wastes into water courses or aquifers.

## Excavation, Overburden/Waste Rock and Tailings Placement Issues

Best practice environmental management of excavation and placement of top soil, overburden/waste rock and tailings not only minimises the environmental impact and facilitates rehabilitation, but can also improve mining efficiency — including improved recovery of the resource and income. Destruction of vegetative cover, whether by siltation, erosion or clearing for mining operations, causes loss of habitat so that ecosystems are degraded and biodiversity is diminished. Environmental best practice includes:

- minimising the clearance of vegetative cover
- retaining top soil so that it can be replaced over mined ground and overburden to support vegetative cover and re-establishment of habitat
- avoidance of ground sluicing unless tailings, waste rock and overburden are impounded in stable forms away from water courses
- keeping mining operations out of active water courses by stream diversion, paddocking or contained dredge ponds.

## Exploration, Definition and Planning for Optimal Recovery of *In situ* Resource

Traditionally, the 'small scale miner's understanding of a mineral resource occurrence is from pre-existing local knowledge. Their information on the quantity, grade, extent, disposition and ground conditions is often minimal — and hence, they are unable to develop the best mine plan and mining method for the local conditions. Mining is commonly by trial and error, resulting in poor recovery, high grading and sterilisation of the unworked resource and compounding environmental impacts. The outcomes are lower net income, resource wastage and greater environmental impact than if the resource was better understood and mining planned from the outset.

Exploration and definition of a mineral resource, together with some basic planning for its extraction, will enable small scale mining to be more profitable, more efficient in resource utilisation and to reduce environmental impacts. However, this involves mapping, sinking pits, trenching, drilling, bulk testing of many samples and geological interpretation. Apart from time-consuming manual crushing of samples, hand-dug pits, washing and pan concentrating this is beyond the means of the individual artisanal miner.

Achieving best practice in environmental management and more efficient production in any mining operation requires planning for all stages, from exploration through to rehabilitation of the disturbed ground. The need for planning is even greater when there are numerous small scale mining operations working the same mineral occurrence. Exploration, definition of the resource, mine planning, and rehabilitation all involve environmental management, which should be integrated for the total resource area or natural landscape unit. This is a function for a government regulatory and advisory agency.

## Impact Assessment, Rehabilitation and Monitoring

Assessing and monitoring environmental impact and rehabilitating ground disturbed by mining are integral parts of medium to large scale mining operations in most countries. These functions are carried out by specialists employed by the corporate or government miner and performance is checked by government agencies. Artisanal miners clearly do not have the means or expertise to carry out these functions in the same way. The EPA wishes to work with governments of developing countries to assist in the development of impact assessment, monitoring and rehabilitation appropriate for small scale mining.

## Labour Migration, Spread of Infectious Diseases, Lawlessness

Where a large number of artisanal miners ōrushō to a mineral field and set up camp, there is no sanitation, infrastructure, stability of family life or established community structure. There is commonly lawlessness, social disruption and spread of infectious diseases. Rushes of artisanal miners are not just an historic feature of the developed new world. Some 500,000 people took part in the 1986 rush in the Philippines. The consequences of such illegal mining can be serious, and as demonstrated at Mt Kare in Papua New Guinea, can result in disastrous conflict with the development of large scale corporate operations.

## Bringing Small Scale Mining into the Formal Economy

Where mining is illegal, the production of high value minerals, such as gemstones and gold, is sold outside the formal economy. It is sold on the black market at a discount to world prices, and often smuggled out of the country of production to the detriment of that country and the miner. In Indonesia, more gold from small scale mining was smuggled out of the country in 1989 than was sold through official channels. Best practice environmental management cannot be adopted for such mining. Legalising mining with licences and facilitating sales through buying facilities that pay world prices has brought an end to black market sales and gold smuggling for Ghana and Zimbabwe. This has brought extra revenue to these countries — $80 million a year in the case of Zimbabwe. It has also opened the door to improved environmental performance by way of obligations, security of tenure and rights that can be attached to a licence.

When artisanal miners are legalised and given security of a licence, their productivity and environmental performance can be improved through positive use of licensing (Burke, 1995). This can include being able to sell to government buying stations at world prices, the availability of equipment to rent at a reasonable fee and access to local mineral processing services. The cost of providing government facilities can be well rewarded. Over a ten-year period Ghana recovered in gold revenue one hundred times the cost of the buying stations (Labonne, 1996). With these services, technical advice and even hands-on training can be provided in the use of equipment and general mining methods by local offices of the government's mineral/mining agency.

Another way of moving small scale and artisanal mining towards best practice is through demonstration recovery plants and mines. In the 1980s, Peter Hancock (of the Centre for Resource and Environmental Studies, Australian National University), on behalf of the Canadian National Research Council, demonstrated the application of best practice small scale alluvial gold mining to miners in Yukon Territory. This included skid and pontoon mounted, trommel and riffle, alluvial gold recovery plants with a closed water circulation system. This mining method is more efficient and has much less environmental impact than the then traditional practice of hydraulicking and ramming giant sluices upstream through sediment in and around active stream beds. Because the method demonstrated produced higher recoveries and enabled more ground to be worked, it was well received and progressively adopted by the local miners. The miners and various government agencies were also pleased to find that they could now avoid massive contamination of streams from the sluicing of overburden and tailings.

## EPA INTEREST AND MANDATE

It is the policy of Australia to assist countries in Asia and the Pacific region in industries and issues where it has expertise. The EPA has a mandate for this on behalf of the government and people of Australia. The exploration, mining and processing of mineral resources is Australia's leading industry. Australian mineral production accounts for Aus$33 billion in export income and approximately 40 per cent of total exports of goods and services. This makes Australia unique, being the only developed country with such a high percentage of mineral industry activity, and demonstrates that a diversified mineral-based economy can support a high standard of living across its population. Australia has become a world leader in mineral exploration, mining and best practice environmental management and is exporting this expertise to assist other countries in sustainable development of their mineral resources.

The Australian government wishes to foster sustainable development, including ongoing improvement of environmental and social conditions for countries in the region. The objective of the EPA is to raise the standard of artisanal and small scale mining through transfer of technology, provision of environmental management expertise and application of practical measures, systems and training. The nature of these measures is determined with representatives of local country government and industry in accordance with country needs and conditions.

## INITIATIVES TAKEN

### Information Modules

The EPA has compiled and distributed a series of information modules, Best Practice Environmental Management in Mining as a series of booklets. These booklets mainly describe large scale, technically advance mining, and use case studies from throughout Australia. The first eleven booklets, listed in Table 1, and a video are now available.

### Table 1: The First Eleven Best Practice Environmental Management in Mining Booklets

*Overview of Best Practice Environmental Management in Mining*
- Overviews the program of best practice environmental management. Outlines the potential problems of mining activities, and some operations recognised for their environmental work. Canvasses the financial benefits of instituting best practice.

*Mine Planning for Environment Protection*
- Explains how planning is the key to identifying and minimising the impacts of mining and how it can help meet community expectations/ aspirations for minimal environmental impacts. It outlines the considerations that shape mining methods and the design of environmental safeguards. These include: air, water and noise quality; transport; biological resources; social and economic factors; surrounding land uses; and heritage places and artefacts.

*Environmental Impact Assessment*
- Introduces the background and purposes of environmental impact assessment (EIA). It covers briefly the legislative requirements within Australia, the key components of EIA, and the different levels of assessment. The relationship of environmental management plans, monitoring and environmental management systems to environmental impact assessment is discussed.

*Community Consultation and Involvement*
- Explains the expectations and needs of communities affected by mining proposals. The processes involved in preparing for the consultation process are discussed in detail and the key community consultation techniques are described. The booklet focuses on a community-centred rather than a project-centred approach to community consultation and involvement.

*Environmental Management Systems*
- Outlines the role and key components of an environmental management system (EMS) as one tool to use in achieving the company's environmental objectives and targets. It explains how to operate, implement and maintain an EMS, from exploration to mine closure.

*Environmental Monitoring and Performance*
- Covers the objectives of monitoring programs; selection of indicators; measurement methods; data collection and analysis; and reporting. Monitoring of water, air, dust, flora and fauna are explained. The linkages between environmental monitoring and performance and environmental auditing and environmental impact assessment predictions are discussed.

*Planning an Environmental Awareness Training Program*
- Explains the importance of planning a work force environmental awareness training program to achieve an enduring and improving environmental culture. Corporate commitment is important to a successful program. A framework is provided which can be used in planning a work force environmental awareness training program and evaluating its success.

**Table 1 contd.**

*Tailings Containment*
- Planning, designing, constructing, operating and monitoring tailings disposal facilities are covered. The factors to consider in selecting suitable sites and the various disposal options for tailings are explained. The monitoring and control methods that can be used to minimise environmental impacts are discussed.

*Rehabilitation and Revegetation*
- The principles and practices of mine rehabilitation are outlined. Particular emphasis is given to the restoration of natural ecosystems, especially the re-establishment of native flora. Topics covered include rehabilitation objectives, soil handling, earthworks, revegetation, soil nutrients, fauna return, maintenance, monitoring and success criteria.

*Environmental Auditing*
- Auditing is shown to be an important tool for any mining operation to measure its performance against current and expected regulatory requirements, improve its credibility with the public, assess its level of risk exposure, and access loan capital. A range of audit types is described and examples given of audit checklists.

*Onshore Exploration*
- Significant environmental damage can result from ground disturbance, clearing of vegetation and careless handling of materials such as drilling fluids, lubricants, fuel, etc. Techniques are described to avoid damage, such as consultation with local people, alternatives to widespread bulldozing, earthworks to minimise erosion, rehabilitation of drill holes, and safe handling of contaminants.

Seven further booklets are in preparation and will cover:

- offshore exploration
- acid mine drainage
- management of hazardous wastes
- mine water management
- cleaner production
- noise and vibration landform design

Additional modules will be produced, resulting in a set of 25–30 booklets.

The booklets are written in simple English language. They are understandable to English speakers with a good general education to high school level (grade 12 in developed countries). They focus on issues, principles, practices and use case studies to illustrate key points. They have been distributed in 60 countries and are being translated into various languages, including Bhasa (Indonesian), Mandarin (Chinese) and Spanish. The first of these have already been launched in Indonesia and have been well received.

### Databases

The EPA has established the "Environet" Australia Database, accessible on the Internet, which provides information on Australia's capabilities in environmental management. This is being further developed with information for miners, training courses and conferences, available environmental expertise, case studies and technical references for specific mining-related issues. The Internet address for "Environet" is http://www.erin.gov.au/net/environet.html

### Training Programs

Training needs for environmental management in mining are being identified in association with the Australian Centre for Minesite Rehabilitation Research. Courses and workshops will use best practice sites as examples. They will target all levels of the work force rather than just focussing on environmental staff. Australia is well located to offer access to training throughout the Asia-Pacific region. The EPA wishes to explore with countries in the region training requirements that will cater for their conditions.

### Target Audience

Achieving best practice environmental management is a partnership affair between the miners, government and the community. The EPA initiatives therefore attempt to involve all these stakeholders to raise their awareness, assist with the initiation of measures to establish best practice environmental management, and achieve a more sustainable development of mineral resources.

### RELEVANCE OF THE EXISTING BEST PRACTICE BOOKLETS

The booklets identify, raise awareness and address the environmental issues for medium to large scale mining. Whilst the general philosophy, many of the issues and some of the prescriptions apply to small scale mining, the EPA is now considering developing products aimed specially at small scale and artisanal mining. It is exploring views and needs to determine what type of product will be most useful. For example, the EPA is working with Indonesia with a view to producing printed materials for artisanal and small scale mining in that country.

In India and other developing countries, as in Indonesia, the EPA wishes to work with government and industry to identify and provide environmental management products that will reduce or avoid the negative effects and facilitate more efficient and economic mining.

## ASSISTANCE OFFERED AND FEED-BACK SOUGHT

The EPA invites developing countries to discuss with it how the Australian mineral sector's experience and the initiatives referred to in this paper and Best Practice booklets can be adapted and used to achieve best practice for small scale and artisanal mining in their country or region. In liaison with those countries, the EPA will then be pleased to adapt the information, according to cultural and literacy situations, into several booklets for small scale and artisanal miners, community and government stakeholders. It will also assist with related training and data sets. Developing country stakeholders' assistance and ideas are sought on how the EPA may work with them to produce a successful product, i.e. a product which describes:

- measures and systems that their miners, communities and governments can afford and that will benefit environmental performance and productivity; and
- measures that communities see as reasonable and which can be made relevant to the small scale and artisanal miners.

## DEVELOPING A STRATEGY TO WORK WITH INDIA AND OTHER DEVELOPING COUNTRIES FOR ENVIRONMENTAL BEST PRACTICE

The work of small scale miners is unspecialised. Environmental performance and practices cannot therefore be neatly parcelled off into tidy specialised compartments — as they are in the existing booklets which are appropriate for large scale mining with its readily available specialist expertise.

The issues to be managed in small scale mining vary widely according to the type of mining, country situation and stakeholders interests. In many cases important issues may not be recognised until awareness of the effects are developed through information, personal experience or changing community values. To help identify the priority issues to be addressed they are grouped and interrelated in various ways by cross-referencing issues with stakeholders, as for example in the hypothetical case for, say, an iron ore quarry as in Table 2.

The stakeholders (eg. artisanal miners, small scale mechanised/ capital intensive miners, government policy-makers, regulators/advisers and community) are listed against the various issues (eg. water use, safety, chemicals, waste materials, noise, dust, etc.) so that the priority concerns can be checked off for any country and stakeholder interest.. Countries are invited to use this approach to help identify the range of relevant issues for small scale and artisanal mining in their country, and help to

Table 2: Identifying the Priority Issues for Country Situations

| Stakeholder / Issue | artisanal miners | mechanised /capital intensive miners | policy makers | regulators/ advisers | community |
|---|---|---|---|---|---|
| water availability/ management | X | X | | | X |
| health and safety | X | X | X | X | X |
| disease/ social unrest/law | | | X | X | X |
| chemical use/ contamination | | | X | X | X |
| vegetation/ biodiversity | | | X | | |
| erosion/ siltation | | | | | X |
| tailings/overburden waste rock placement | | | X | X | |
| resource utilisation/ sterilisation | | | X | X | |
| productivity | X | X | | | |
| noise | | X | | | X |
| dust | | X | | | X |

determine what the most useful products would be to address these issues.

## EXAMPLES OF ACHIEVABLE OUTCOMES

Positive outcomes for small scale and artisanal mining have been achieved by legalising mining through licences issued to mine within a specified area. Zimbabwe and Ghana provide positive incentives by buying gold at world market prices and at the same time can impose reasonable performance conditions on the licensee. Licensed small mine sites can then be both assisted and controlled. The following are examples of potentially achievable outcomes:

- simple mine rehabilitation and environmental management plans to be completed and followed for each small mine site and to comply with regional requirements
- linkage of licensing to inspection, access to mining equipment, inspection, advice and training
- best practice demonstration sites and plant, such as simple trommel and riffle plants for alluvial mining with continuous rehabilitation and closed system water management
- Infrastructure for small scale mining areas, providing efficient crushing, grinding, concentrating and amalgam retorting facilities.

## REFERENCES

Burke, Gill 1995. "Policies for Small Scale Mining: the need for Integration", *Conference Proceedings, Mining and Mineral Resources Policy Issues in Asia-Pacific*, pp. 103-106, Australian National University, Canberra

Labonne, Beatrice 1996. "Artisanal mining: an economic stepping stone for women" *Natural Resources Forum* Vol 20, No 2, pp. 117-122.

Li Peiji and Li Guangwei 1996. "The Development and Prospect of Small Scale Mining in China", unpublished Paper of Chinese Mining Association, Beijing.

Noetstaller, Richard 1995. "Historical perspective and key issues of artisanal mining": Keynote Speech at International Round Table on Artisanal Mining, May 1995, The World Bank, Washington, DC.

# INTEGRATED ENVIRONMENTAL MANAGEMENT IN SMALL SCALE MINING — A BOLIVIAN EXPERIENCE

*Guillermo Cortez*
Integrated Environmental Management in SSM, Bolivia
(Swiss Agency for Development and Cooperation, S.D.C.)


The environmental issue is nowadays in Bolivia one of the most important aspects to consider in any project and especially in mining projects.

However, while the large State owned mines and the private Medium Scale Mining companies have cooperation and financing from organizations like the World Bank or from private investors, the Small Scale Mining sector does not have any cooperation nor funds to solve the pollution problems caused by their operations.

The Environment Law approved in 1992 and its regulations promulgated in April this year need to be disseminated and explained to the miners and especially with reference to EIA's and what is known in Bolivia as the "Environmental Manifesto".

This situation moved the Swiss Agency for Development S.D.C. to help in the solution of environmental problems originated in the Small Scale Mining sector which embraces both the mining cooperatives and the small scale miners. A German consulting group Project/Consult GmbH was selected to design and execute a program aimed to cooperate the small scale mining sector in the solution of those environmental problems.

The program named "Integrated Environmental Management in the Small Scale Mining", has the Ministry of Environment as counterpart, its objective is to explain to the miners that the mining operations should be carried out integrated with environmental protection measure, if so, they will not be closed down and will participate in the country's sustainable development.

Two years and a half later, (operations started April 1994), the program has executed a series of activities in the training, education and dissemination of environmental principles and regulations applied to mining operations in the small scale mining sector.

Two areas have been considered the most important because the high degree of pollution they cause:

1. Mercury pollution related to gold mining in both alluvial and hard rock or vein type mining.
2. Water pollution caused by Acid Mine Drainage from recent and old operations, dumping of waste to the rivers causing siltation, and effluents emissions from mills and treatment plants.

The use of mercury was and still is irrational, miners ignore its toxic effects, they amalgam by hand and volatilize it in open air.

The program carried out the following:

- Optimization of mercury use in gold recovery, avoiding its use in open circle.
- Dissemination of techniques and equipment to use mercury adequately.
- Make the miners conscious about proper handling of mercury to protect health and environment.

With reference to water and soil pollution, the worst point was Potosi city where nearly 40 beneficiation plants discharge tailings containing lead, zinc, silver with pyrite and other sulphides into the rivers, polluting considerably agricultural soils, fauna and flora and the Pilcomayo river which is an international course.

The program has contracted a U.K. consulting group to prepare the feasibility study for a tailings pond which it is expected will be commisioned in the near future.

Seminars, courses and other meetings as well as periodicals and magazines funded by the program have been used to disseminate as much information as possible on environment and mining. Practical demonstrations in the field have shown to the miners gold recovery techniques especially fine gold which up to the present has been lost to the rivers nearby.

Some gold recovery equipment and machines, like retorts and amalgamating drums, have been sold to the miners at cost prices. Some plants have been modified to avoid use of mercury in open circle and improve gold and mercury recovery.

The program has signed agreements with different institutions NGO's, public and private organizations, representatives of mining cooperatives to solve different problems in the environmental field.

Several university students are working on Thesis that will help to solve some specific problems.

Results so far are encouraging and the change in mentality of leaders and miners with respect to environmental protection is the best justification to continue the program for another term.

# ENVIRONMENTAL AND SOCIAL CONSIDERATIONS DURING ABANDONMENT OF SMALL SCALE UNDERGROUND MINING OPERATIONS: CASE STUDY OF A COPPER MINE

A. Santha Ram* and A.N. Bose**
*Senior Mining Geologist, Mining Research Cell
**Controller of Mines (Central Zone)
Indian Bureau of Mines, Nagpur, India

## INTRODUCTION

In Indian conditions, the environmental standards related to mining industry have changed over the past few decades. It is now recognized by the operating mining companies, the local government and the public, that the environmental damage that can result from mining operations, can be regulated under several environmental laws to ensure that they cause least possible damager and abatement measures will be required to protect the environment.

One of the largest environmental remedial costs involved are with the state owned and large private owned mines. Earlier, the concept of reclamation was not seriously considered during the project planning stage, due to long gestation and operating periods.

As the mine workings become uneconomic, the mine operators are forced to cease the operations to bring the life cycle of a mine to an abandonment. Thus abandoned or inactive mines pose a variety of environmental problems and generation of data on environmental factors is very important for assessing the likely hazard potential. The realistic detrimental factors include mine type, minerals produced, host rock, beneficiation method, period of active operation, etc. It is now recognized the world over that prior to the abandonment of the mining operations, the following information should be available:

a. A brief description of the project with technical details: (Mining method, beneficiation method, tailing disposal, etc.)
b. The relevant data required to identify and assess major effects.
c. The possible measures envisaged to minimize or avoid significant likely damage, if any.

In general, as on date there are no specific stipulations laid down by the Government agencies for mine abandonment plans or details of restoration and rehabilitation purposes. The EIA (Environmental Impact Assessment) and EMP (Environmental Management Plan) studies are usually addressed to upcoming mining projects, but not related to their closures. Thus, the mine owners wish to eliminate the future liabilities at the earliest, prior to abandonment, and to satisfy the controlling authorities that the abandonment and rehabilitation have been carried out in accordance with a mutually agreed plan.

## ABANDONMENT PLANS FOR SMALL/MEDIUM UNDERGROUND MINES

The underground mine includes temporary, and permanent workings such as adits, inclines, shafts, haulage ways. The permanent workings include shafts, underground utilities. The waste rock generated during mining is either filled within the stopes or some times brought up to the surface for disposal.

The waste rocks sometimes contain residual minerals, which may contribute to acid generation, which may further cause acid mine drainage and corrosion of mine equipment.

The programme of mine abandonment and the methods and means of site rehabilitation and restoration now form an integral part of planning. Thus the ultimate goal is to create conditions after the cessation of mining operations that allow a minimal risk to the environment and the local populace.

## REQUIREMENT OF MINE ABANDONMENT UNDER RULE 23(2) OF MINERAL CONSERVATION AND DEVELOPMENT RULES (MCDR) 1988

Before the mine reaches to stage of abandonment, the mine management is required to file an application to the Indian Bureau of Mines (IBM) furnishing the following information and plans/sections:

1. Duly filled prescribed form-E to designated officers at least 90 days before the intended date of such abandonment.

2. True copies of Plans and Sections of the mine on a scale not less than 1:1000, setting forth accurately the work done in the mine up to the time of submission of the notice including measures envisaged for protection of the abandoned mine or part thereof, approaches thereto and the environmental.

The Controlling authorities will inspect the mine and related infrastructure facilities, if required, to evaluate the adequacy or otherwise of the measures from mineral conservation and environmental point of view. This is usually accomplished by conducting special studies and/or inspections which may involve systematic sampling and analysis of mine water, mill water, tailing water, effluents, stabilisation of tailing dams, revegetation programmes, re-estimation of existing ore reserves based on techno-economic considerations and examination of relevant records/analysis results etc. The deficiencies, if any, observed will be intimated to the mine management for further action before a formal permission in writing for abandonment of the mine is granted by authorised officer of the IBM who may prohibit abandonment or allow it to be done with such conditions as may be specified in the order.

Rule-34 of these rules stipulates further that every holder of a mining lease shall undertake a phased restoration, reclamation of areas affected by mining operations and shall complete this work before conclusion of such operations and abandonment of the mine. Compliance of the provisions of this rule will also be required before abandonment of a mine.

## MINE ABANDONMENT: A CASE STUDY

A small-scale underground copper mine of 100 tonne per day capacity with matching beneficiation facilities is located in Rajasthan and 230 kms, south west of Delhi, the capital of India. The physiography of the area is marked by two long and high ridges separated by a major and wide valley.

The deposit was initially explored by the IBM by drilling from surface and underground. Exploratory mining was done from 1956 to 1963, and as a result two potential mineable blocks were developed. IBM estimated about 4,00,000 tons of ore with 2.35% Cu grade under proved category and 1,35,000 tons of ore with 2.80% Cu under probable category. The total lease area is 914 hectares.

Production from the mine commenced in 1974, envisaging ore production for more than 10 years. But after 6 years of successful mining within the mine block, there was a sharp decline in ore production, due to non-availability of mineable ore. Further, the deposit had been re-evaluated in 1980, to examine the economic viability for 100 tpd ore production at mill feed grade of 2.14% Cu, with a cut off grade of 0.5% and 2 m was of mineable width.

The project started in February, 1972 and out of the mineable ore reserves of 2,80,000 tons approximately 2,00,000 tons were mined till 1978. However, with further exploration and proving of additional reserves, and the mining operations continued till middle of 1994.

Exploration has been carried out subsequently to establish additional reserves. However, as further occurrence of ore could not be established, there was no alternative left but to close down the mine, due to depletion of reserves, the mining operations become uneconomical. The mine had a small township providing housing for their managers, workers and other functional buildings like manager's office, vocational training centre, canteen, bank and a primary school.

When the decision for closure was taken, there were about 38,000 tons of geological reserves left out in the mine. Out of this, only about 13,000 tons were mineable. Rest of the reserves were unmineable due to various reasons viz. ore body being beyond stoping limit, less width, some ore to be left in the pillars to ensure safety of mine workings etc. In order to recover this tonnage of mineable ore reserves, an additional 190 m of mine development would have been required, which included 40 m of mine development in waste rock, for which alone an additional expenditure of Rs. 20,00,000 (US$ 66,000 approx.) would have been incurred.

## Mining Geology

The copper mineralisation is confined to the rocks of Precambrian age. Chalcopyrite is the primary mineral associated with pyrite and pyrrohotite. Phyllites are the host rocks for copper mineralisation. The ore bodies are tabular to lensoid in shape, plunging at 15°C to 20°C in the south west direction. There are two distinct parallel main lodes. The footwall lode is represented by a number of ore shoots, exhibiting an enechelon pattern. Most of the lenses in the footwall lode have been mined out. The hanging wall lode has been exposed by systematic exploration, but no significant reserves could be proved.

## Mining Operations

The mine has been developed as an underground mine to produce 100–150 tpd copper ore by shrinkage method of stoping. The access to the mine is by two adits and one vertical shaft. These adits were located in the mine block and the vertical shaft located in the nallah block. About 13 stope blocks have been developed which were subsequently stoped out, and about 30,000 tons of geological reserves were left unmined, due to uneconomic conditions.

## Concentrator Plant

A 100 tpd concentrator plant is located near to the mine site. Based on the detailed laboratory investigation of this sulphide ore, IBM designed the process flow sheet for improving the recovery efficiency and to minimise the tailing losses. At present, the beneficiation plant is totally dismantled and some of the more expensive equipment have been removed. Since the generation of tailing depends upon the nature of ore minerals milled, therefore, a brief description of ore characteristics and process practiced have been discussed.

Chalcopyrite is the primary mineral associated with pyrite and pyrrohotite. The associated minerals are cubanite (Cu, $Fe_2$, $S_2$), Magnetite ($Fe_3O_4$), Pentalandite (FeNiS) and Sphalerite (ZnS).

*Process Description*

The process involved primary and secondary crushing followed by grinding and flotation. The ROM ore is crushed from 250 mm to 13 mm in two stages. The secondary crushing is done by Symons cone crusher to –13 mm material and stored in fine ore bin. The ore is ground to about 65% –200 mesh. The pulp is fed to a cyclone, operating in close circuit with ball mill at 250% circulating load. The flotation cells consist of 250 cft each for roughing and 4 number of cells of 10 cft capacity for cleaning. The thickened copper concentrate is filtered in disc filter to about 85% solids. The grade of concentrate averages 18% Cu. The tailings contained around 0.1% Cu and 60 ppm of cobalt and 120 ppm of Ni. The tailing pond occupied 2,80,000 cubic metres. The tailing effluents are properly treated before final discharge.

## ENVIRONMENTAL CONSIDERATIONS

The environmental aspects examined prior to the granting permission for abandonment of the mine included the following:

1. Study of surface and underground environment.
2. Study of concentrator plant, tailing disposal, effluents, discharge of water in nearby aquifers, wells, streams, etc.
3. Water Management:
   a. Installation of site drainage
   b. Monitoring surface run-off
   c. Monitoring mine water.
4. Site rehabilitation.
5. Vegetation of tailing and waste dumps.

Among the above-mentioned parameters, the study of surface and underground environment is very important for deriving useful information regarding its impact. In the milling process, tailing disposal depends on the nature of ROM (Run-of-Mine) ore type, operational capacity, quantum and nature of chemical reagents used and total recovery percent of water in the concentrate.

Tailings from the concentrator plant had been dumped over an area measuring approximately 300 m x 160 m, located on the north east. Before dumping the tailings, the dump site was made impervious to arrest groundwater contamination. A seasonal stream flows towards the northern side of this dump.

Further north of the tailing dump, the area comprise of few agricultural lands and habitations. During monsoon, surface run off from the nearby hills flows over this tailing dump, forming deep gullies. As a result of this process, some quantity of tailings were transported to longer distances in the form of slimes.

The slimes could pollute a nearby perennial stream. The airborne dust generated from the tailing dump is transported to nearby agricultural fields due to high wind velocities, particularly during summer even though fairly good consolidation had taken place due to the presence of pyrrohotite in the tailings.

## ABATEMENT MEASURES ADOPTED

In the case of this mine, the abatement measures already adopted by the management includes monitoring the quality of drinking water from the nearby wells. The quality of water from the wells for drinking purpose is reproduced in Table 1.

Periodic analysis of treated effluent discharge water from the tailing dump is carried out to ascertain transport of metallic ions into the stream. The chemical analysis of effluent water has revealed that the quality of water is within permissible limits. The results are summarised in Table 2.

The mine water has been analysed for various chemical constituents. The analysis has revealed that the quality of mine water is within the permissible limits.

The chemical analysis of water samples collected from the underground is summarised in Table 3.

There was no appreciable change in the quality of ground water. Water samples from the area were analysed frequently and regularly by different agencies.

## Table 1: Analysis Results of Drinking Water

| Sl.No. | Parameters | Limits as per IS-10500-1983 | 1986 | 1988 | 1990 | 1993 | 1994 | 1995 |
|---|---|---|---|---|---|---|---|---|
| 1. | pH | 6.5 - 8.5 | 7.29 | 7.73 | 8.06 | 7.58 | 7.6 | 7.5 |
| 2. | Total dissolved solids | 300 mg/lit | 964 | 742 | 748 | 1200 | 1400 | 1050 |
| 3. | P-Alkalinity | -- | N.T. | 6 | 13 | -- | 15 | 5 |
| 4. | Mg-Alkalinity | -- | 470 | 343 | 328 | 300 | 450 | 110 |
| 5. | Total Hardness | 600 mg/lit | 480 | 380 | 423 | 602 | 600 | 590 |
| 6. | Calcium hardness (Mg/l) | 75 as Ca | 250 | 200 | 222 | 332 | 300 | 380 |
| 7. | Magnesium Hardness (mg/lit) | 30 as Mg | 230 | 180 | 201 | 270 | 300 | 210 |
| 8. | Silica as $SiO_2$ (mg/lit) | -- | 15 | 33 | 36 | 38 | 40 | 50 |
| 9. | Iron as Fe | 0.3 mg/lit | N.T. | N.T. | N.T. | N.T. | N.T. | N.T. |
| 10. | Sulphate as $SO_4$ | 400 mg/lit | 205 | 118 | 141 | 345 | 350 | 226 |
| 11. | Fluoride as F | 0.6 to 1.2 mg/lit | 0.90 | 0.71 | 0.78 | 1.12 | 1.17 | N.D. |
| 12. | Copper as Cu | 0.05 mg/lit | N.T. | N.T. | N.T. | N.T. | N.T. | N.T. |
| 13. | Phosphate as P | -- | 0.02 | 0.03 | N.T. | N.T. | N.T. | N.T. |
| 14. | Chloride as Cl | 250 mg/lit | 140 | 95 | 87.5 | 175 | 190 | 170 |
| 15. | Zinc as Zn | 5 mg/lit | N.T. | 0.55 | N.T. | N.T. | N.T. | N.T. |
| 16. | Lead as Pb | 0.1 mg/lit | N.T. | N.T. | N.T. | N.T. | N.T. | N.T. |
| 17. | Nickel as Ni | -- | N.T. | N.T. | N.T. | N.T. | N.T. | N.T. |
| 18. | T.R.C. | 0.2 mg/lit | N.T. | N.T. | N.T. | N.T. | N.T. | N.T. |
| 19. | Nitrate as $NO_2$ | 45 mg/lit | N.T. | N.T. | N.T. | N.T. | N.T. | N.T. |
| 20. | Arsenic | | | | | | | 0.1 |
| 21. | C.O.D. | | | | | | | 20 |
| 22. | Colour | | | | | | | < 5 |

N.T.: Not traceable.  N.D.: Not determined

## Table 2: Analysis of Effluent Water

| Sl. No | Parameters | Limits as per IS-2490 Part 1-1981 | 1986 | 1988 | 1992 | 1993 |
|---|---|---|---|---|---|---|
| 1. | pH | 5.5-9.0 | 7.97 | 7.54 | 7.92 | 7.9 |
| 2. | Colour | Hazen 75 | < 5 | < 5 | < 5 | < 5 |
| 3. | T.S. Solids | 100 mg/lit | 54 | 33.6 | 49.5 | 29 |
| 4. | Sulphide as S | 2 mg/lit | 0.11 | 0.05 | 0.075 | 0.01 |
| 5. | Fluoride as F | 2 mg/lit | 1.1 | 1.32 | 1.01 | 0.01 |
| 6. | Copper as Cu | 3 mg/li | N.T. | N.T. | N.T. | N.T. |
| 7. | Lead as Pb | 0.1 mg/lit | N.T. | N.T. | N.T. | N.T. |
| 8. | Nickel as Ni | 5 mg/lit | N.T. | N.T. | N.T. | N.T. |
| 9. | Zinc as Zn | 5mg/lit | 0.1 | 0.05 | 0.06 | N.T. |
| 10. | Phosphate as $PO_4$ | 15 mg/lit | 0.03 | N.T. | 0.04 | N.T. |
| 11. | Iron as Fe | 0.3 mg/lit | N.T. | N.T. | N.T. | N.T. |
| 12. | Arsenic as As | 0.2 mg/lit | 0.003 | 0.11 | N.T. | N.T. |
| 13. | C.O.D. | 250 mg/lit | 21.3 | 25.3 | 23 | 8 |
| 14. | Oil & Grease | 10 mg/lit | N.T. | N.T. | N.T. | N.T. |
| 15. | Dissolved Oxygen | -- | 7.0 | 4.3 | 6.0 | 6.3 |

Table 3: Analysis of Mine Water

| Sl. No. | Parameters | Analysis |
|---|---|---|
| 1. | pH | 1.5 |
| 2. | T.D. Solids | 1600 mg/lit |
| 3. | Total Hardness | 400 mg/lit |
| 4. | Silica | 60 mg/lit |
| 5. | Iron as Fe | 39 mg/lit |
| 6. | sulphate | 1012 mg/lit |
| 7. | Fluoride | 4.0 mg/lit |
| 8. | Chloride | 15.0 mg/lit |
| 9. | Copper | 5.5 mg/lit |
| 10. | Zinc | 2.3 mg/lit |
| 11. | Nickel | 1.3 mg/lit |
| 12. | Nitrate as $NO_2$ | N.T. |

## Solid-Waste Management

The waste rock produced during mining operations was brought to the surface through an adit and dumped over a nearby hill slope. The quantum of waste dumped till abandonment so far was around 75,000 cu.m. which needed stabilisation and vegetation. A stonewall about 0.5 m high had been constructed to prevent any wash off from this dump.

Plantation over waste dump and tailing dump had commenced for stabilising the existing dumps. Around 1600 nos. of saplings were planted on the waste dump and around 20,000 of saplings including "Prosphis julia flora" were planted over the tailing pond and in rest of the lease area.

## SOCIO-ECONOMIC CONSIDERATIONS

Mines are quite often located in areas where they constitute the main economic resource. The closure of such operations therefore has significant socio-economic impact. The abandonment of this mine and concentrator resulted in an socio-economic impact directly on the work force, management as well to the supporting small-scale industries in and around neighbouring villages. Since the small mining industry at such a remote location has developed a community around itself, therefore its socioeconomic effect cannot be evaluated in terms of compensation. However, the management had given careful consideration before abandonment to take care of the welfare of the work force and their families.

Prior to abandonment, the mine and mill had a workforce of 241 workmen and 15 executives. Out of which 119 workmen and 3 executives opted for voluntary retirement scheme from the company. The remaining workmen and executives opted for transfer to another nearby sister unit of the same company.

After the abandonment of the mining operations the tailing dumps were stabilised by growing plants/trees over and on the sides of the tailing dumps and retaining walls have been constructed to minimise the run-off tailings. Around 1500 saplings have been planted on the tailing dumps during 1994–95, the survival rate being 50–60%. During the year 1995–96, around 700 more saplings were planted.

As the project was commissioned, the socio-economic environment of the area changed considerably. The basic amenities provided by the project during its life cycle of operations included:

1. Educational facilities: A Senior Secondary school was started as such facilities was available earlier.
2. Medical facilities: A dispensary was started for providing the basic medical amenities.
3. Communication facilities: Post office and telephone facilities were provided.
4. Transportation: Transport, construction of approach roads to nearby villages.
5. Drinking water: free supply of drinking water to the nearby villages.
6. Power supply: Power supply was made available in most of the villages.

## ABATEMENT MEASURES SUGGESTED

The abandonment of the mine was allowed subject to certain conditions which included:

1. A retaining wall to be built by the company between the public road and the northern side of the tailing dump to prevent or minimise any wash off from the tailing dump due to the presence of seasonal stream, agricultural fields and human settlement.
2. The tailing dump and waste dump should be stabilised by vegetation and the plantation as already carried out or proposed to be carried out need to be monitored by the company till such time the mining lease is to be surrendered to the State Government.
3. The quality of ground water should be monitored in the vicinity of the tailing dump, during pre- and post-monsoon period.
4. Monitoring of air borne dust from the tailing dump and its likely impact in the neighbouring agricultural fields need be carried out till the surrender of the lease.

## CONCLUSIONS

The abandonment plans for mines need to cover underground and surface facilities, infrastructure available at the mine site, water management, site rehabilitation which include spoil and tailing dump areas and socio-economics. The ultimate objective is to create conditions after the cessation of mining activities which will pose minimal effects to the environment and local populace. Site inspection followed by limited time bound investigations were conducted prior to the abandonment of mining and milling operations.

Based on the finding, the mine management was advised to adopt suitable measures for the minimal damage of the environment, before the surrender of lease to the State Government. The steps include monitoring of water quality from wells, quality of effluents that were discharged into natural water courses, protection against acid mine drainage by sealing off the underground entries. Steps were also suggested for minimising the air-borne dust from the tailing dump by planting *Phosophis julia* flora. Due to the total stoppage of mining and milling operations some overflow of water from mine and plant is expected, and necessary abatement measures adopted which, however, were not adequate. Necessary permission for the abandonment of the mine was granted after a careful consideration of all these factors which are likely to contribute to possible environmental impacts.

## ACKNOWLEDGEMENT

The authors acknowledge with thanks the kind permission granted by the Controller General, Indian Bureau of Mines, Nagpur. Valuable data have been generated by the team of IBM officers (Ajmer Region) during abandonment stage in the form of special studies. The authors wish to thank Shri L.K. Trehan, General Manager (Mines), Khetri Copper Complex, Hindustan Copper Limited for providing valuable technical information related to the project.

### REFERENCE

Parmar R.L., Taluja S.C., and Kanhayalal (1994). "Report on Special Study on Kho-Dariba Copper Mine, Hindustan Copper Ltd., IBM Report (unpublished).

# DEVELOPMENT OF AN ALGORITHM FOR INTEGRATED ENVIRONMENTAL MANAGEMENT INFORMATION SYSTEM FOR SMALL SCALE OPEN CAST MINES OF HIMALAYAN REGION

*A.K. Soni and A. Swarup*
Scientist, Central Mining Research Institute, Regional Centre
Roorkee 247 667, U.P.

## INTRODUCTION

Exploitation of mineral resources by surface mining methods have adverse effects on the environment. In ecologically fragile areas of Himalaya, environment and its management assume special importance both as a subject and object on account of its ecological fragility and environmental sensitivity for both biotic and abiotic life. Computer as a tool for conversion of raw input to a usable output have led to development of INFORMATION SYSTEMS (IS). Small-scale surface mines which are in vogue in Himalayan region can thus find tremendous application of such IS for the flow of information from the data generating agencies to the real users in the field in a ready to use form. Implementation of such information for environmental management of these mine unit(s) will find better utility due to the fact that these mines are small in size and located in remote areas.

## WHAT IS INTEGRATED ENVIRONMENTAL MANAGEMENT INFORMATION SYSTEM

Information on environmental science and technology and management of environmental information is very vast, multilingual and widely scattered. In totality, it is difficult to approach to all knowledge base due to the well- known and obvious facts. Manual methods constrain the full use of data (or information) on account of infrastructural and financial constraints. Therefore, such information remain concealed and the user industry remain devoid of harnessing its benefit. The task of using data or information collected at a given point of time or space can find utility in context to a different place and time if a knowledge base is created. This knowledge base or information network can then be linked to a networking system electronically for local, regional, national and international need. Figure 1 indicates basics of Integrated Environmental Management Information System (INEMIS). It comprises of two main parts viz information and management of data pertaining to environment. Thus, defining critically, "Integrated Environmental Management Information System is a tool which essentially help in facilitating flow of environmental information to the users at a desired time, which is varied in form and uneven in quality. It also helps in analysis of a given environmental data to design solution to a environmental problem."

The word "integrated" used in context to environmental management can be described as that scientific approach to rational mountain development which causes least harmful environmental effects of any developmental activity whether mining or otherwise and is a multidisciplinary strategy involving the co-operation of local population.

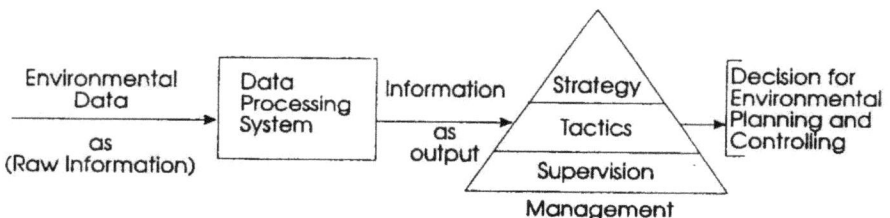

Figure 1: **Basics of INEMIS**

## DEVELOPMENT OF ALGORITHM

In Himalaya large number of small scale mines are operative which accounts for nearly 40% of the production. Figure 2 presents comparative profiles generally considered typical for small and large mining operations. As opposed to large scale mining, small mines are the preferred alternative as their requirements in terms of reserves are less, implementation time and initial investment are small whereas skills and infrastructure requirement are moderate and employment per unit output is high. In terms of environmental attributes and management of these attributes (Air environment, Water environment and Land environment) it is comparatively less cumbersome for mines of small scale.

Conventionally, these environmental attributes can be evaluated, compared or calculated by manual means. Using computer this task can be simplified and as a first step towards this is the development of logical steps or algorithm to be followed. This, in turn, can be developed as a

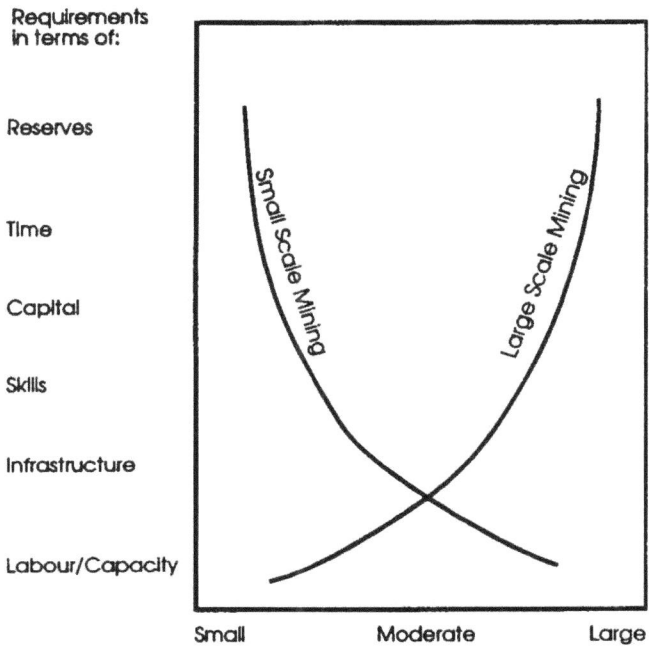

**Figure 2: Small vs. Large-scale Mining Comparative Profiles
(after Noetstaller, 1991)**

total software package for application to small-scale mines for a particular area in sight. The overall framework for this can be described below.

SMNEMIS: INEMIS developed for particular application to Small mines is named as SMNEMIS—Small Mines Integrated Environment Management Information System. It encompasses two components—

Part I : Environmental Information.
Part II : Environmental Management.

As the name envisages the part I is a database of environmental information whereas part II facilitates the use of database of part I for determining the status of environment. Model format of database for various information such as atmosphere/climate, species, wildlife, land categories, bibliography etc is to be prepared. Two such model format for atmospheric and species database are given in Annexure I.

The algorithm for part II is given in Figure 3(a) and can be divided into three main modules as given below beside input and output data module. The Programme Data Control File (PDCF) is responsible for calculation and logical conclusion for air, water and land modules. A separate flow sheet is provided for PDCF in Figure 3(b).

| Input Data Module |
|---|
| Module (I) : AE Module |
| Module (II) : WE Module |
| Module (III): LE Module |
| Output Data Module |

(Here abbreviation AE, WE, LE refers to air environment, water environment and land environment respectively).

Modules I & II determine the qualitative status of air and water environment respectively by comparing the actual field values of the various air and water parameters than those of standard values prescribed by the regulatory agencies. In respect of air environment, Table 1 indicates the ambient air quality standards as prescribed by Central Pollution Control Board (CPCB) for sensitive, industrial and residential areas as applicable to the mine sites. For qualitative assessment of water quality standards laid down by Environment Protection Rules (EPR-1986) as given in Table 2 may be followed.

The percentage of land degradation in a mining area considering mine as equivalent to a WATERSHED can be determined using remote sensing and GIS techniques which is not described here. The land quality assessment according to the capability, the land possess can be done as follows:

Step 1: Select one area/slope of the particular watershed and check whether that particular slope is rainfed or lying on which side of the Sun.
Step 2: Input salient points of watershed description data after determining the area of the land to be managed.
Step 3: Evaluate the land using empirical approach (land capability) based on actual field data/observations obtained from the field.
Step 4: Categorise the land evaluated in step 3.
Step 5: Check and judge the suitability of evaluated land qualitatively, whether it can be converted to agricultural landuse or other non-agricultural landuse and verify the qualitative results than those of quantitative results evaluated in steps 3 and 4.
Step 6: Design the benches/terraces scientifically.
Step 7: Involve local people and on the basis of socio-economic condition of inhabitants find their interest of potential landuse. Compare the present landuse pattern from the potential landuse to arrive at a sound decision based on practical considerations.
Step 8: Suggest land management strategy for different land types. All land can be classified in eight sub-categories or classess from Class I to VIII as given in Annexure II.

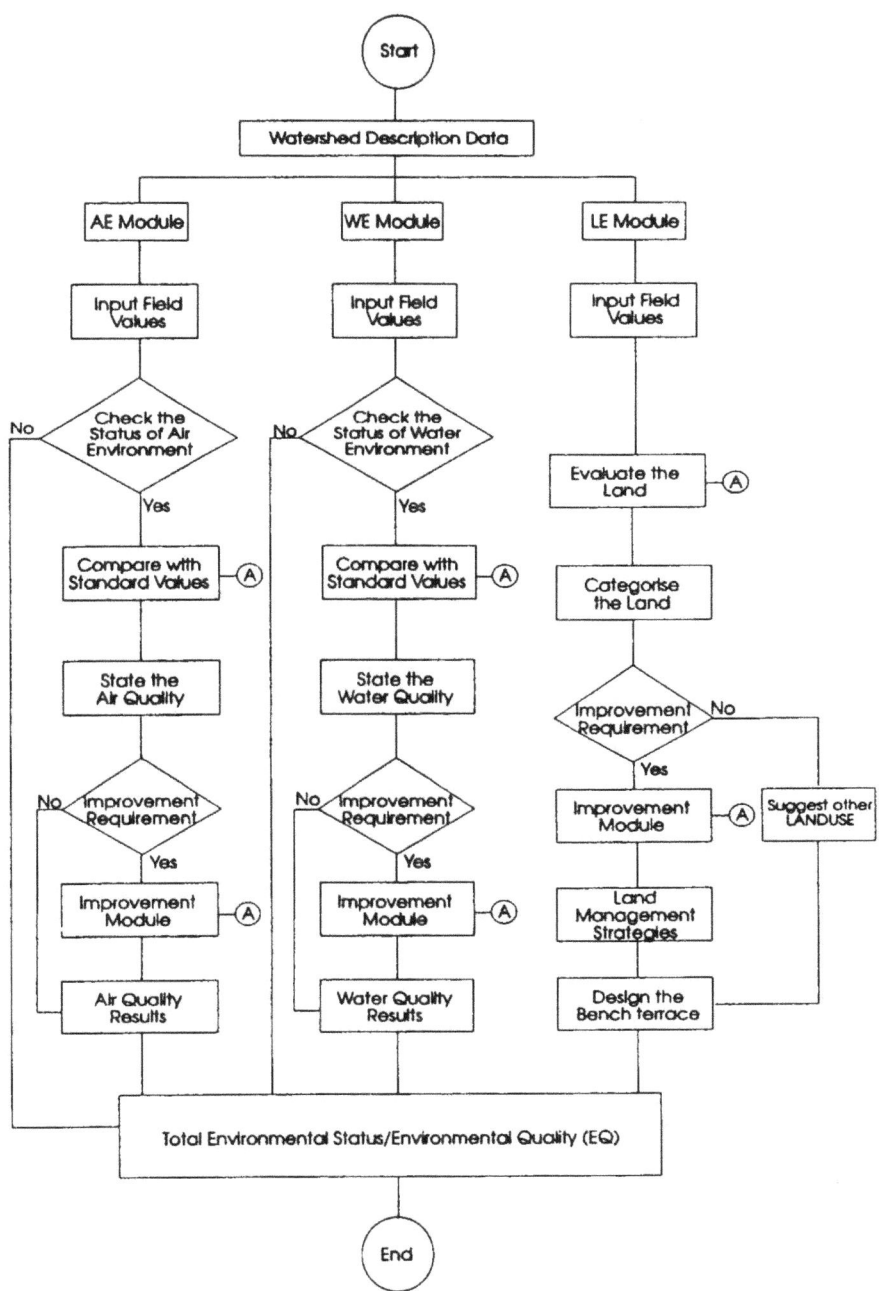

Figure 3(a): Flow Chart of SMNEMIS.

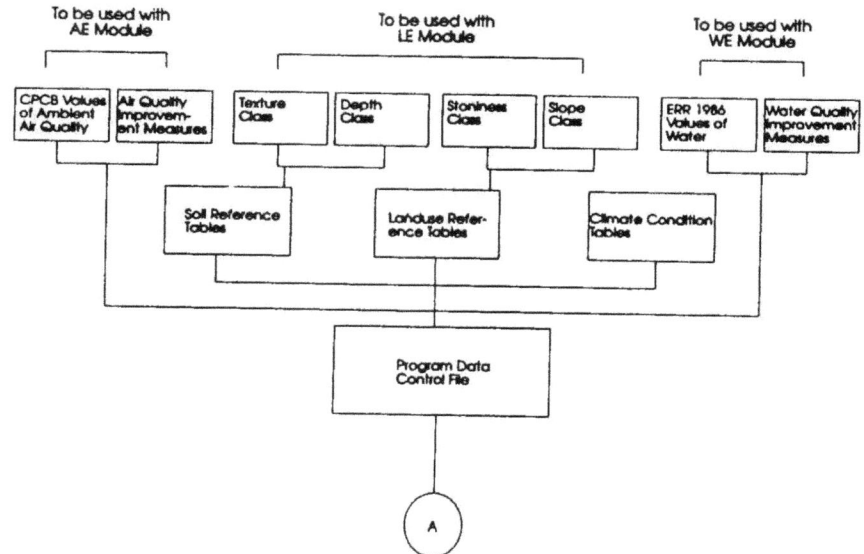

**Figure 3(b): Program Data Control File (PDCF).**

**Table 1: Ambient Air Quality Standard**

| S. No. | Category | SPM | $SO_2$ | CO | $NO_x$ |
|---|---|---|---|---|---|
| 1. | Environmentally sensitive areas | 100 | 30 | 1000 | 30 |
| 2. | Industrial areas and Mixed uses | 500 | 120 | 5000 | 120 |
| 3. | Residential and rural areas | 200 | 80 | 2000 | 80 |

Source : CPCB publication.
Note: All values are expressed in $g/m^3$ unless stated.

The selected input parameters for land quality assesment are given in Table 3.

## Advantage of SMNEMIS

(1) The most vital effect of environmental degradation in a hill area is particularly in terms of land degradation (creation of ugly scars on the hill slopes). SMNEMIS helps to quantify it according to the capability the land possess.

(2) With the advent of remote sensing and Geographic Information System (GIS), it is possible to digitize the land use data and integrate both the technologies, which would facilitate higher accuracy for land management as far as a local or smaller area of watershed is concerned. SMNEMIS can be used in the present form for this purpose and is thus compatible for use with GIS.

(3) It is ready to use and is user-friendly.

## Table 2: General Standards for Discharge of Effluents
### (Schedule II, Rule 3, Environment Protection Rules 1986)

| S. No. Parameters | Standards | | | |
|---|---|---|---|---|
| | Inland surface water | Public sewers | Land for irrigation | Marine coastal areas |
| 1. Colour and odour | See note 1 | – | See note 1 | See note 1 |
| 2. Suspended solids mg/l. max | 100 | 600 | 200 | (a) For processs waste water=100 (b) For cooling water effluent -10% above total suspended matter of inffluent cooling water. |
| 3. Particle size of suspended solids | shall pass 850 micron IS sieve | – | – | (a) Floatable solids max 3 mm. (b) Settleable solids max 850 microns |
| 4. Dissolved solids (inorganic)mg/l. max | 2100 | 2100 | 2100 | – |
| 5. pH value | 5.5–9.0 | 5.5–9.0 | 5.5–9.0 | 5.5–9.0 |
| 6. Temperature in degree C Max | Shall not exceed 40 in the and section point of the stream point within 15 metres downstream from the effluent outlet | 45 at the point | – | 45 at the point of discharge. |
| 7. Oil and Grease mg/l. max | 10 | 20 | 10 | 20 |
| 8. Total residual chlorine mg/l. max | 1.0 | – | – | 1.0 |
| 9. Ammonical Nitrogen (as N) mg/l. max | 50 | 50 | – | 50 |
| 10. Total Kjeldahl nitrogen (as N) mg/l. max | 100 | – | – | 100 |
| 11. Free Ammonia(as $NH_3$) mg/l. max | 5.0 | – | – | 5.0 |

*contd.*

Table 2 contd.

| S. No. | Parameters | Standards | | | |
|---|---|---|---|---|---|
| | | Inland surface water | Public sewers | Land for irrigation | Marine coastal areas |
| 12. | Biochemical Oxygen Demand (5 days at 20°C). max mg/l. max | 30 | 350 | 100 | 100 |
| 13. | Chemical Oxygen Demand mg/l. max | 250 | – | – | 250 |
| 14. | Arsenic(as As) mg/l. max | 0.2 | 0.2 | 0.2 | 0.2 |
| 15. | Mercury (as Hg) mg/l. max | 0.01 | 0.01 | – | 0.01 |
| 16. | Lead (as Pb) mg/l. max | 0.1 | 1.0 | – | 1.0 |
| 17. | Cadmium(as Cd) mg/l. max | 2.0 | 1.0 | – | 2.0 |
| 18. | Hexavalent Chromium (as Cr + 6).mg/l. max | 0.1 | 2.0 | – | 1.0 |
| 19. | Total Chromium (as Cr).mg/L | max | 2.0 | 2.0 | – |
| 20. | Copper(as Cu) mg/l. max | 3.0 | 3.0 | – | 3.0 |
| 21. | Zinc(as Zn) mg/l. max | 5.0 | 15 | – | 15.0 |
| 22. | Selenium (as Se) mg/l. max | 0.05 | 0.05 | – | 0.05 |
| 23. | Nickel(as Ni) mg/l. max | 3.0 | 3.0 | – | 5.0 |
| 24. | Boron(as B) mg/l. max | 2.0 | 2.0 | 2.0 | – |
| 25. | Percent Sodium max | – | 60 | 60 | – |
| 26. | Residual Sodium Carbonate mg/l. max | – | – | 5.0 | – |
| 27. | Cyanide (as CN) mg/l. max | 0.2 | 2.0 | 0.2 | 0.2 |
| 28. | Chloride(as Cl) mg/l. max | 1000 | 1000 | 6000 | – |
| 29. | Fluoride(as F) mg/l. max | 2.0 | 15 | – | 15 |

| | | | | |
|---|---|---|---|---|
| 30. Dissolvedphosphate (as P) mg/l. max | 5.0 | — | — | — |
| 31. Sulphate (as $SO_4$) mg/l. max | 1000 | 1000 | 1000 | — |
| 32. Sulphide phosphate (as P) mg/l. max | 2.0 | — | — | 5.0 |
| 33. Pesticides | Absent | Absent | Absent | Absent |
| 34. Phenolic Compounds | 1.0 | 5.0 | — | 5.0 |
| 35. Radioactive Materials: | | | | |
| a) Alpha emitter MC/ml. max | $10^7$ | $10^8$ | $10^8$ | $10^7$ |
| b) Beta emitter μC/ml. max | $10^6$ | $10^5$ | $10^7$ | $10^6$ |

N.B. (i) All efforts should be made to remove colour and unpleasant odour as far as practicable.

(ii) The standards mentioned here shall apply to all the effluents discharged such as industrial mining and mineral processing activities, municipal sewage etc.

(iii) These standards shall not apply to those industries for which standards have been notified by the central government vide S.O. 844(E) dated Nov. 18, 1986, S.O. 393 (E) dated 16th April 1987, S.O. 443 (E) dated the 28th April 1987 and S.O. 64 (E) dated the 18th January, 1988. This notification shall cease to apply with regard to a particular industry when industry specific standards are notified for that industry.

(iv) Schedule II is added by notification no GSR 919 (E) dated 12.9.1988 published in Gazette of India, Extra, Part II, Sec. 3 (i) dated 12/9/1988.

**Table 3: Selected Input Parameters**

| S. No. | Parameters | Values | Remarks |
|---|---|---|---|
| 1. | Soil pH | 0–14 | This parameter defines alkalinity or acidity of soil and is an important parameter for selection of plant species. |
| 2. | Soil Texture | Ratio of percentage of solid space to percentage of pore space = 1 (ideal case) | This parameter gives picture about the soil ability to hold water in the inner pores and gives percentage of sand to silt to clay. It is a supportive parameter and can be determined by nomograph or feel method. |
| 3. | Depth of soil cover | 0 to > 90 cm | Soil depth is not a limiting factor. |
| 4. | Stoniness | < 0–1 to > 4 or (< 10% soil to 80% soil) | It is an important parameter for mining land. Larger the stone percentage less will be the soil and accordingly land will be less suitable for productive purposes. |
| 5. | Slope | 0° to > 33° | Degree of slopiness of the ground decide the design of benches/terraces to optimise the net land area available. |
| 6. | Climate Classes | C1 to C5 classes | C1 = Humid with well distributed rainfall<br>C2 = Humid climate with occasional dry spells<br>C3 = Sub-Humid crop yields frequently reduced by droughts<br>C4 = Semi Arid<br>C5 = Snowy and Arid type climate |

## CONCLUSION

On the basis of the developed algorithm as given in the paper and following the steps described a fully menu-driven computer package having the facilities of determining environmental parameters, calculation of environmental parameters, searching, appending, editing and browsing of environmental related information can be developed to cater to the need of users in mineral industry in general and Himalayan region in particular.

### REFERENCES

1. Handbook of Agriculture (1992). ICAR, New Delhi, pp 92–120.
2. Gore and Stubbe 1984). Computer and Information System, Chapter 3, pp 37–52, Mc-Graw-Hill International edition.
3. Noetstaller R. (1991). "Small Scale Mining: Practises, Policies, Perspectives", Editor: Ajay K. Ghose, Oxford & IBH Publishing Company Private Ltd., New Delhi, pp 3–10.
4. Soni A.K. (1996). Unpublished Draft of Ph. D. Thesis.

## ANNEXURE - I
### Part I : Environmental Information

### (A) MODEL FORMAT OF ATMOSPHERIC DATABASE

Name of Mine        :                Watershed Name :
Name of Station    :
   Latitude      -
   Longitude    -
   Elevation     -
Date of Recording -
of information

                            Month
                 Jan Feb Mar Apr May June .... Dec.

(1) Precipitation/Rainfall
(2) Snowfall
(3) Temperature in °C
Daytime       Maximum in °C
                Minimum in °C
Nighttime     Maximum in °C
                Minimum in °C
(4) Wind Speed

### (B) MODEL FORMAT OF SPECIES DATABASE

Name of Mine                          :
Watershed Name              :
Date of Registering information :
Name of Species            :
Family of Species          :
Details of Location— Region    :
                         Location  :
                         Altitude   :
                         Category  :
                         Type      :
Lifeform                              :
Rare/Endangered          :

## ANNEXURE - II

*Class I Lands (Green colour):-* have generally deep soil of high fertility, consisting of mostly best quality loam of high moisture retention capacity and rich in humus. These are found in either flat areas or on gentle slopes by and large free from any hazards of erosion and water-logging. The physical environment favours their maximum utilisation and are best suited for agriculture uses.

*Class II Lands (Yellow colour):-* are characterised by coarse-textured soils, liable to water-logging, (occasional damaging overflow), suffer from slight to moderate erosion, less than ideal soil depth, wetness which can be corrected by drainage but existing permanently as a moderate limitation, slight to moderate salinity and a slight climatic limitations on soil use and management.

*Class III Lands (Red colour):-* are the lands with a marked degree of slope and have soil of coarse texture. Both erosion and water-logging impose serious limitations on its use. Soils categories in class III land reduces the choice of plants and require special conservation practices. Thus, class III land has one or more of the following factors.

(i) Moderately sloping land.
(ii) Moderately susceptible to water or wind erosion.
(iii) Frequent overflow accompanied with some crop damage.
(iv) Very slow permeability of the sub-soil.
(v) Wetness or continuing water-logging after drainage.
(vi) Shallow soil depth up to the bed-rock, hard-pan or clay-pan which limits the rooting-zone and the water storage.
(vii) Low moisture-holding capacity.
(viii) Moderate salinity or sodium content.
(ix) Moderate climatic limitation.

The soils can be used for raising cultivated crops, pastures, forests and wild-life food and cover.

*Class IV (Blue colour):-* These class of lands are those lands which are characterised by one or more of the following features
(a) Steep slopes or slopes increases rapidly.
(b) Severe susceptibility to water and wind erosion
(c) Severe effect of past erosion
(d) Shallow soils
(e) Low moisture holding capacity
(f) Frequent overflow with severe crop damage

(g) Excessive wetness with a continuing hazard of water-logging after drainage
(h) severe salinity
(i) moderately adverse climate.

These lands can be used for either crops purposes, pastures, forests or wild-life food and cover.

*Class V (Dark Green or Uncoloured):-* Such land is nearly level and is not subject to more than slight wind to water erosion. Cultivation is not feasible because of one or more limitations, such as overflow, stoniness, wetness or severe climate. Examples of class V land are: (i) Lowlands subject to frequent overflows which prevent the normal production of cultivated crops, (ii) nearly level soils with growing season that prevents the normal production of cultivated crops, (iii) the level or nearly level, stony or rocky soils and (iv) ponded areas where drainage for cultivated crops is not feasible but where soils are suitable for grasses or trees. Soils in class V are not suitable for raising cultivated crops, but are suitable for perennial vegatation (grazing and forestry, with few or no limitations). Pastures can be improved, and benefits from proper management can be expected. Physical conditions of the soils are such that it is practicable to apply pasture improvements, if needed, such as seeding, liming, fertilizing, and water control with contour furrows, drainage ditches, diversions of water spreaders. Soils in class V have little or no erosion hazard, but have other limitations, the removal of which is not practicable. They are used largely for pastures, forests and wild-life food and cover.

*Class VI (Orange colour):-* Lands in class VI have limitations which cannot be corrected, such as (i) a steep slope, (ii) very severe erosion hazard, (iii) very severe effect of past erosion, (iv) stoniness, (v) shallow rooting-zone, (vi) excessive wetness or overflow, (vii) low moisture capacity, (viii) salinity or sodium content and (ix) severe climate. Soils in this class are subject to moderate lilmitations under grazing or forestry use. They are generally unsuitable for cultivation and limit their use largely to pastures or forests or wild-life and cover.

*Class VII (Brown colour):-* Lands in class VII have very severe limitations that make them unsuitable for cultivation and that restrict their use largely to grazing or forestation or wild-life food and cover. The soils in this class are subject to severe limitations or hazards under either grazing or forestry use. The physical condition of soils is such that it is not practicable to adopt pasture improvements and water-control practices. Soil restrictions are severe than those in the case of class VI soils.

*Class VIII (Purple colour):-* Land forms and soils in class VIII have limitations that preclude their use for commercial plant/trees production and restrict their use to recreation, wild-life food and cover or to water-supply, watershed protection or for aesthetic purposes. Bad lands, rock outcrops, sandy beaches, marshes, deserts, river wash, mine tailings and other nearly barren lands are included in class VIII.

Significant return or site benefits from soils and land forms in class VIII cannot be expected from management of crops, grasses or trees, although indirect benefits form wild-life, watershed protection or recreation may be possible.

Note: The colours indicated here are used on various land maps for different purposes and are followed as standard.

# SECTION 3

# Technical Developments

# BUILDING A BETTER SLUICE BOX

*Don Stewart*
International Development Technologies Centre, Faculty of Engineering,
The University of Melbourne, Australia

## INTRODUCTION

The sluice box for the recovery of heavy minerals, notably gold, from placer deposits is one of the earliest forms of mineral processing equipment, being at least two thousand years old. It has changed very little in that time. The sluice boxes described by Pliny (circa AD 70) and Agricola (1556) in "De Re Metallica" are essentially the same as sluice boxes in use today. While simple sluice boxes have largely been superseded in modern mechanised mineral processing operations they are still one of the most widely used pieces of mineral processing equipment because of their use by small-scale local miners in many developing countries as well as by small operators and hobby miners in developed countries. The differences in design of sluice boxes are only minor but the method of operation can vary considerably. It is a tribute to the efficacy of the sluice box that it operates in a manner which appears to satisfy its users under a variety of conditions. Of course, it is also true that these operators do not have any alternative simple low cost technique available and that in most cases they do not have any information on recoveries in their operations. Data on this aspect of sluice box operation is scant but it would appear that recovery of gold of less than 100 microns in a simple sluice box is poor, certainly less than 50% (Wang and Poling, 1981; Fricker, 1984).

Given its antiquity and its ubiquity it would at first appear surprising that the principles of operation of the simple sluice box are not well understood and have not been the subject of much scientific study. Classic texts on ore dressing (e.g. Taggart, 1945) give a great deal of information about the use of sluice boxes in different mines but the evaluation of boxes is mainly anecdotal. Modern textbooks on mineral processing (e.g. Kelly and Spottiswood, 1982; Burt, 1984) treat the topic of sluice boxes but give no further insights into understanding the principles of operations. A more recent review of "The Principles of Sluicing" gives some useful insights but deals mainly with modern systems such as the pinched sluice or the Reichert cone (Sivamohan and Forssberg, 1985). A previous paper (Stewart, 1986) discusses the physical processes potentially operating in a simple sluice box but did not have sufficient experimental data to assess their importance.

Three recent papers (Subasinghe, 1990; Subasinghe and Boardia, 1990; Subasinghe, 1991) have, however, dealt with the simple sluice and have provided a useful model of the passage of material through the box. In this paper we have adopted a somewhat different approach but we believe that those papers and the current work should be seen as complementary. The studies and Subasinghe and Bordia have a similar rationale to this current study; both their work and this current study have arisen out of a desire to improve the operations of small-scale gold miners working in developing countries who use the sluice box as their sole, or primary method of gold recovery (Stewart and Gayap, 1989; Stewart, 1990). In many cases these miners are only achieving low recoveries of gold.

This paper deals firstly, with the basic principles of how a simple sluice box recovers heavy minerals and then the application of this understanding to the development of an improved design for a simple sluice box.

## EXPERIMENTAL

### Apparatus

The experimental studies were carried out in a perspex sluice box (3.5 x 0.1m) connected to a variable speed recirculating slurry pump. The sluice box was fitted with a calibrated notched weir for flow measurements and the whole apparatus could be tilted to vary the slope of the box. The sluice box was fitted with various designs of riffle as required by the experiment. A video camera was used to tape dye injection and flow pattern studies so that they could be studied in detail.

### Materials

The gangue material used in the studies was either a screened natural river gravel with the screen analysis shown in Table 1 or closely sized gravels or mixtures of these. The gravels used were 2000 micron (+1700–2360), 1500 micron (+1700–2360), 1500 micron (+1180–1700) and 750 micron (+600–850).

The "heavy mineral" in the study was either iron (S.G., 7.9), copper (S.G., 8.9) or lead (S.G., 11.3) of sizes 350 micron (+300–425), 200 micron (+150–212) or 75 micron (+53–106).

Table 1. Screen analysis of river gravel

| Size (micron) | Percent retained |
|---|---|
| + 9,500 – 16,000 | 19 |
| + 4,750 – 9,500 | 27 |
| + 2,360 – 4,750 | 19 |
| + 1,000 – 2,360 | 15 |
| +600 – 1000 | 8 |
| +212 – 600 | 9 |
| +106 – 212 | 2 |
| –106 | 1 |

## Method

In the heavy mineral recovery experiments the water flow in the box was adjusted to the required flow rate and feed was added to the sluice box upstream of the first riffle. Unless otherwise stated, the feed consisted of 1000 g of gravel and 100 g of heavy mineral (if that was added). Flow was continued for a period of time (20 minutes unless otherwise stated) and then the water flow was stopped, material removed from the various riffle compartments and analysed for gangue and heavy mineral. Separation of gangue and heavy mineral was normally by screening or magnetic separation. In its "standard" configuration the sluice box contained six riffles each 28 mm high and 197 mm apart. Unless otherwise stated, the slope of the box was maintained at 4 degrees to the horizontal.

Calculations were made of the average mass per riffle compartment, excluding material upstream of the first riffle and downstream of the final riffle. Grade and cumulative recovery calculations were also made. These include the material upstream of the first riffle, although a point corresponding to this material was not calculated.

## RESULTS AND DISCUSSION

### Gravel Only Runs

The typical flow pattern in a simple sluice box is shown in Figure 1(a). An eddy is formed downstream of an obstruction such as a riffle and it is at this position that material collects at high flow rates. At low flow rates material will collect upstream of the riffles because at low flow rates movement through the box is by a flowing film mechanism (Stewart, 1986) or a rolling/sliding mechanism (Bordia and Subasinghe,1990; Bordia, 1990) and the riffles impede movement by this mechanism. Using the modified Shields equation Subasinghe and Bordia (1990) have derived an equation defining the onset of turbulent suspension or the high flow rate regime when material collects downstream of the riffles. Figure 2 shows our

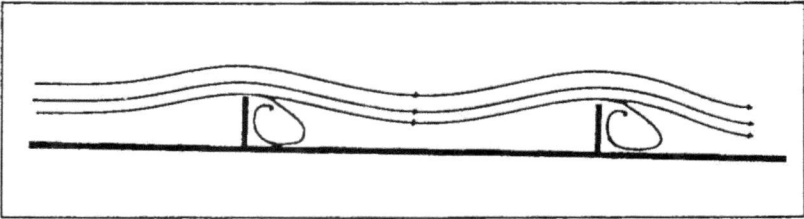

(a) Flow patterns in standard sluice box.

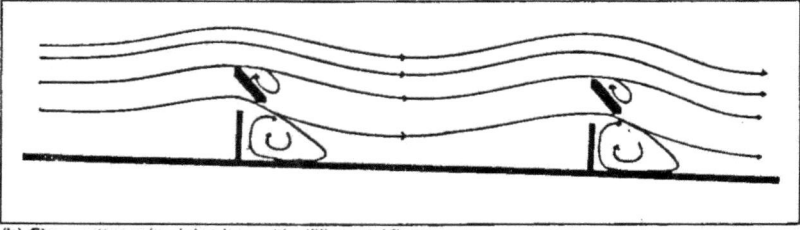

(b) Flow patterns in sluice box with riffles and flaps

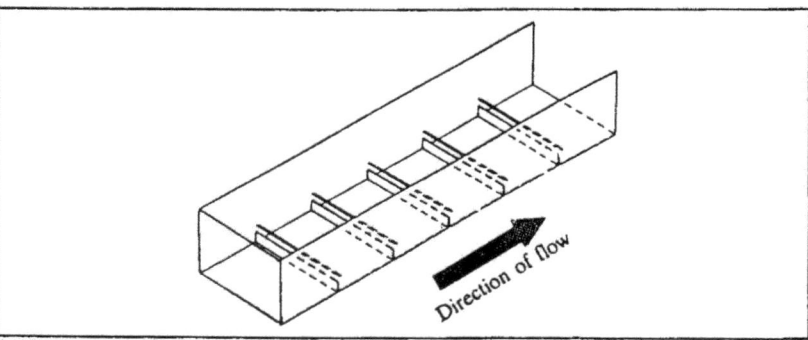

(c) Arrangement of riffles and flaps in modified sluice box.

**Figure 1. Sluice box systems**

results for the average mass per riffle compartment for gravel of different sizes and different flow rates. It shows that once the transition to turbulent suspension has occurred increase in flow does not change the quantity of solid retained except that at very high flow rates there is a slight increase in average mass per riffle compartment. This behaviour is consistent and is characteristic of the sluice box. It is because at high flow rates the flow pattern shown in Figure 1(a) can no longer develop so larger dead zones are formed in every second riffle compartment and thus alternate riffle compartments contain more material. The flow rate when this occurs depends, of course, on the particular sluice box geometry but a sluice box

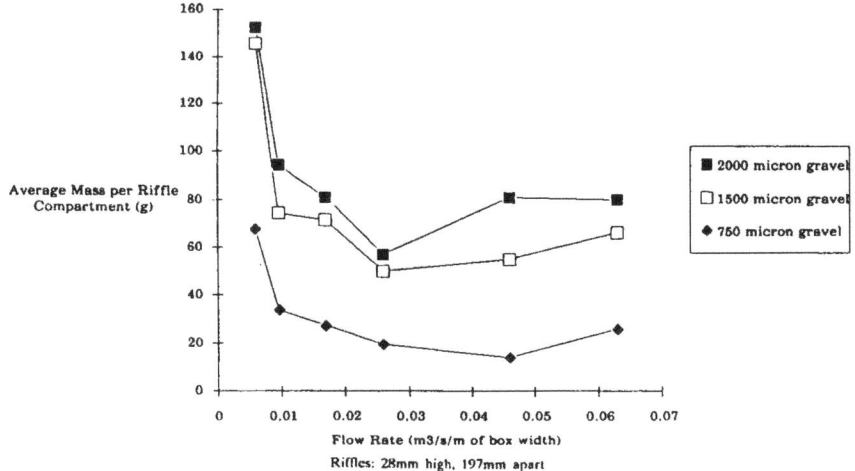

Figure 2. Flow rate vs av. mass per riffle compartment

exhibiting this phenomenon is not operating efficiently. This shows the need to design a sluice box with the conditions of operation in mind. Most of the data collected in this study were for sluice box behaviour in the turbulent suspension regime because this is the normal area for practical operation of the sluice box, although some miners in developing countries work in the flowing film area (Stewart, 1986). There are, of course, transition flow rates where material collects both upstream and downstream of the riffles.

Our studies of the change in the mass of material contained in a sluice box over time show that, irrespective of flow rate, once material has settled in the quiescent zones it is only slowly eroded away by the flow of water, even if there are no further additions of solid feed.

### Gravel and/or Heavy Mineral Runs

When the feed to a sluice box is homogeneous particulate material the question of to what extent that material is retained in the box is determined by the hydrological properties of that material which in turn are determined by particle size, shape and density. When the feed is not homogeneous then the situation is not so simple. In Figures 3, 4 and 5 we see typical results indicating the effect of the presence of gravel on the movement and retention of heavy mineral in the sluice box. In Figure 3 we see that at flows greater than 0.02 $m^3/s/m$ of box width no 75 micron iron is retained within in our sluice box when it is the only constituent of the feed. There is a marked increase in the amount retained when the feed also contains gravel. Figure 4 shows a similar effect for 175 micron

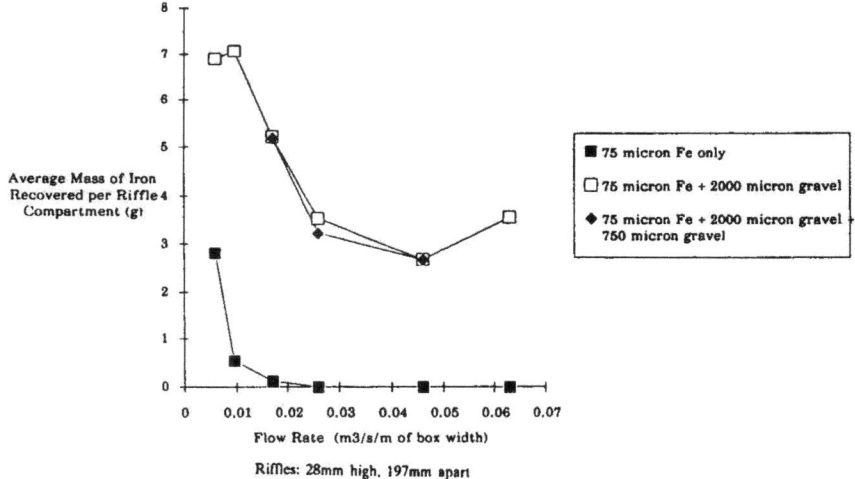

Figure 3. Effect of gravel on heavy mineral recovery (75 micron iron)

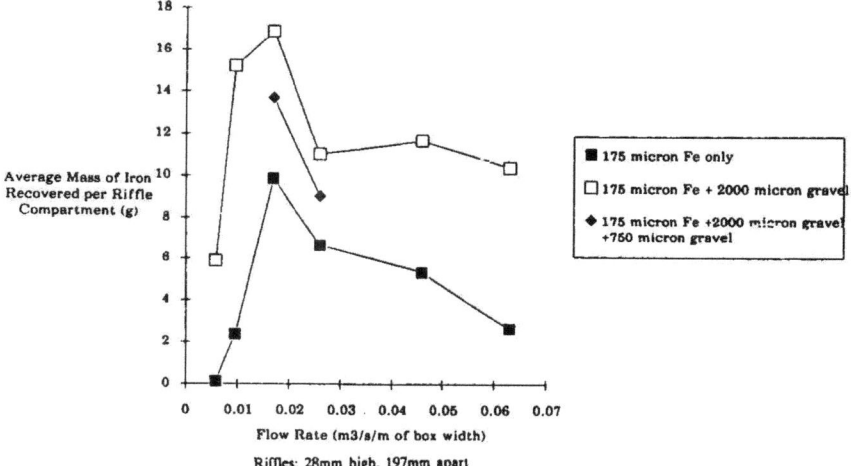

Figure 4. Effect of gravel on heavy mineral recovery (175 micron iron)

iron, although in this case the effect is not quite so marked. Figure 5 shows a different effect. In this case the 350 micron copper will not move into the box at very low flow rates but is trapped and carried through with the gravel under these conditions.

It can be postulated that the gravel shields the fine heavy mineral particles from being washed through the sluice box by providing a quiescent zone in the interstices between the gravel grains. If this is the case we would expect the packing of the gravel to have an influence since there must be voids in the gravel matrix for this mechanism to function.

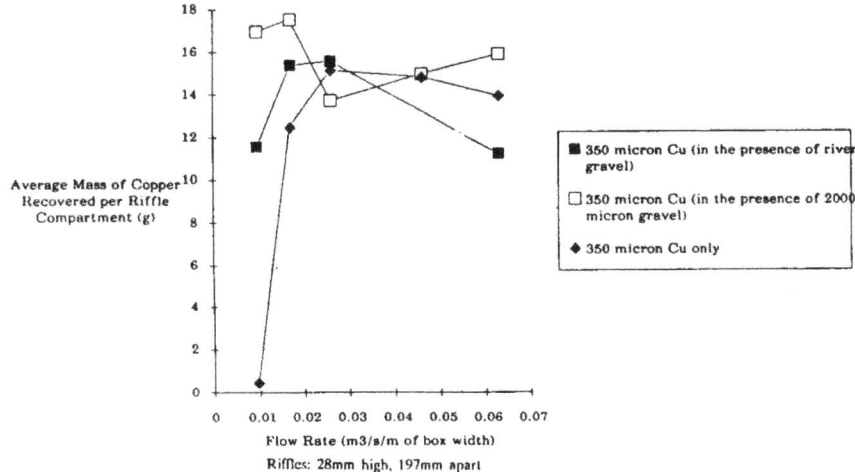

**Figure 5.** Effect of gravel on heavy mineral recovery (350 micron copper)

Some evidence for the reduction of the effectiveness of this mechanism when fine gravel is also present is seen in Figure 4 when 750 micron gravel presumably occupies some of the voids which would otherwise be available for the 175 micron iron. An alternative model for this effect is given by Mayer (1964) in his description of the jigging process. In Mayer's terms the presence of finer gravel would increase the bulk density of the gangue and thereby reduce the driving force for stratification. As expected, Figure 3 shows much less effect from the presence of 750 micron gravel on the trapping of 75 micron iron. We would expect that, in a properly functioning sluice box, fine gangue would be swept out of the box thus leaving voids for the heavy mineral to shelter in. Figure 5 indicates that this mechanism does not seem to be important with larger heavy mineral grains.

Figure 6 is an example of how these mechanisms effect grade and recovery. As we might expect the highest recoveries are found with only large gravel present and the lowest recoveries when only fine gravel is present. It is interesting to see the good recovery achieved when the feed is natural river gravel, showing that these shielding mechanisms are working effectively in this case. The variations in grade shown in Figure 6 can be explained by the effect on weight of gravel retained of a combination of the hydrological and packing properties of the gravel samples used.

## Changes in Sluice Box Slope

In many respects the effect of increasing box slope is similar to the effect of increasing flow rate in terms of aiding the passage of material through

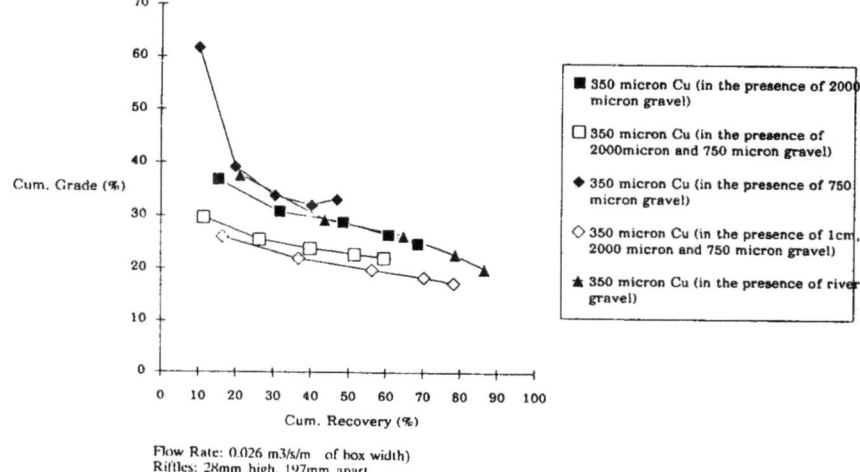

**Figure 6. Effect of gravel on heavy mineral recovery (350 micron copper) and product grade**

the box. However, increasing the slope of the box also has the effect of reducing the dead space in the box, and hence the volume of solids retained in the box. This effect can be seen in Figure 7. This reduction in volume of solids retained tends to cause a reduction in recovery.

## Changes in Riffle Placement

The simplest changes which can be made to sluice box geometry are to change the spacing and height of the riffles. Lowering the height of the riffles in the box has the following effects:

(a) The depth of flow in the box is reduced and so at any given volumetric flow rate the instantaneous fluid velocity is increased. This means turbulent flow is developed at lower volumetric flow rates.

(b) The dead space for collection of gravel, and hence heavy mineral is reduced.

(c) The dead space will be subjected to greater turbulence.

Both effects (a) and (b) can be seen in Figure 8 while Figure 9 shows how recovery of heavy mineral is reduced by reducing the riffle height. Effect (c) above means that greater movement in the settled (dead space) material causes an increase in grade and this can also be seen in Figures 9 and 10. The reduced recovery due to lower riffle height can be compensated by increasing the number of riffles (i.e. reducing their spacing) and so, as can be seen in Figures 9 and 10, reducing both the height and spacing results in a higher grade without any great decrease in recovery. The limitation to proceeding down this road is that for effective box opera-

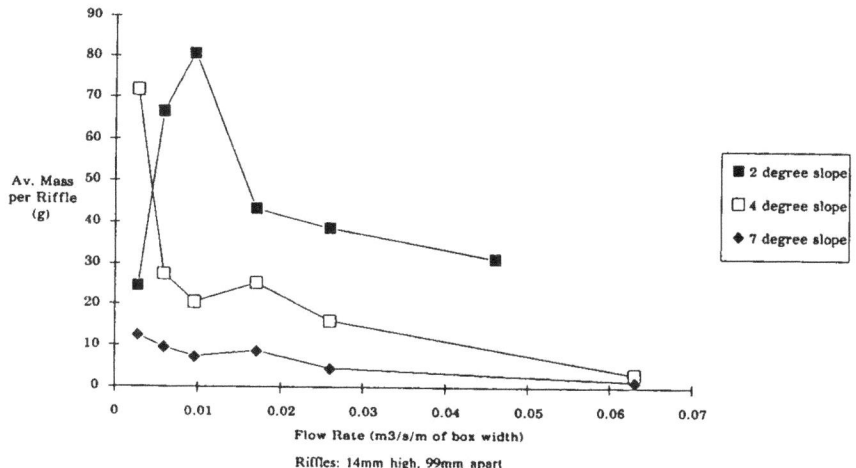

**Figure 7. Effect of slope on av. mass per riffle compartment (2000 micron gravel, riffles at half height, half spacing)**

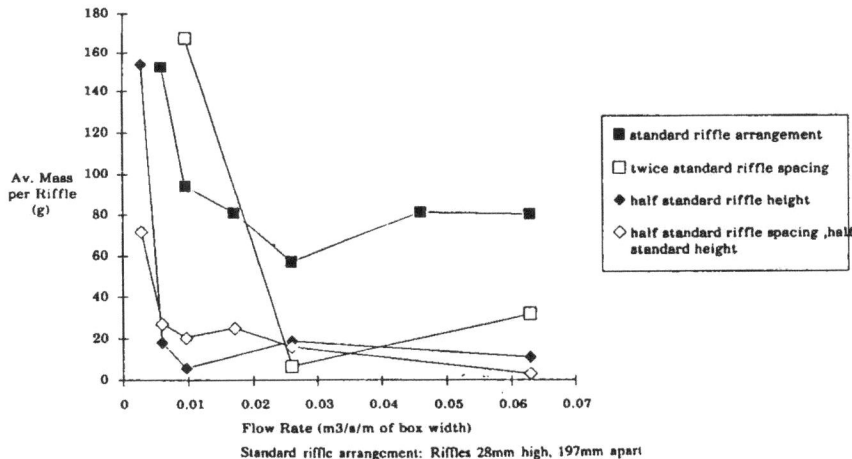

**Figure 8. Effect of riffle height and spacing on av. mass per riffle compartment (2000 micron gravel)**

tion we need to develop the flow pattern shown in Figure 1(a) between each riffle so the riffles need to be sufficiently widely spaced for this to occur. The higher the flow rate the wider the spacing needs to be. The larger the feed size the higher the flow rate needed so in qualitative terms we can say that for fine material low, closely spaced riffles are required and flow rate, riffle height and spacing all need to be increased for larger gravel.

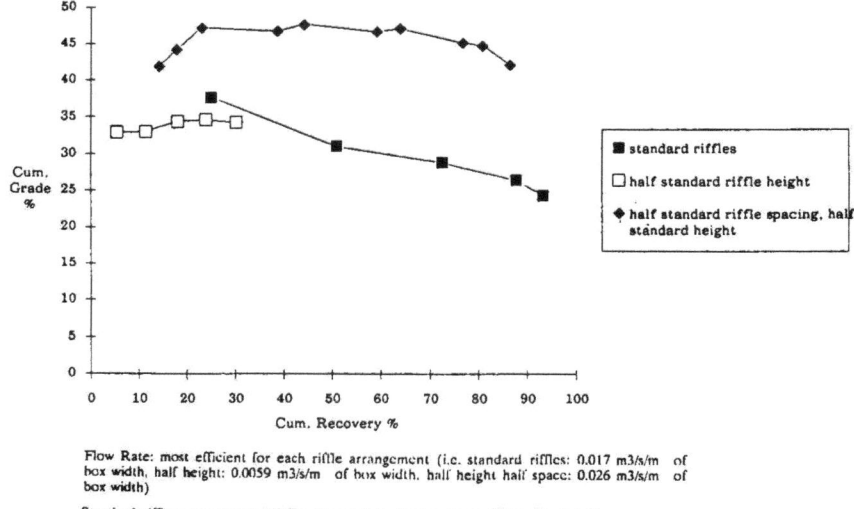

Flow Rate: most efficient for each riffle arrangement (i.e. standard riffles: 0.017 m3/s/m of box width, half height: 0.0059 m3/s/m of box width, half height half space: 0.026 m3/s/m of box width)

Standard riffle arrangement: Riffles 28mm high, 197mm apart, Flaps 20mm wide

**Figure 9. Effect of riffle height and spacing on recovery and grade of heavy mineral (350 micron copper, 2000 micron gravel)**

Figure 10 shows that if the heavy mineral consists largely of fine material then a smaller riffle height, closer spacing and lower flow rates are the more effective conditions for high recovery.

## Changes in Riffle Geometry

There is a wealth of folklore in the industry about riffle geometry with testimonials to the advantages of various arrangements such as Hungarian riffles — placed facing both downstream and upstream and a great variety of other riffles shapes (Taggart, 1945). While there is not a lot of hard data to support many of the claims made it is true that riffle geometry does have an effect on sluice box performance because it modifies flow patterns in the sluice box.

Based on our understanding of the mode of operation of the sluice box we have approached riffle geometry modification with the objective of maximising the effectiveness of the dead zone in trapping heavy mineral. This means that the dead zone should contain a stable deposit of gravel which does not contain fine gangue so that there is space in the interstices for the heavy mineral. There should also be some movement in this bed of gravel to assist in the movement of the heavy mineral into the bed but the movement should not be too great otherwise the protection afforded by the gravel will be largely negated. With these criteria in mind two different approaches were investigated.

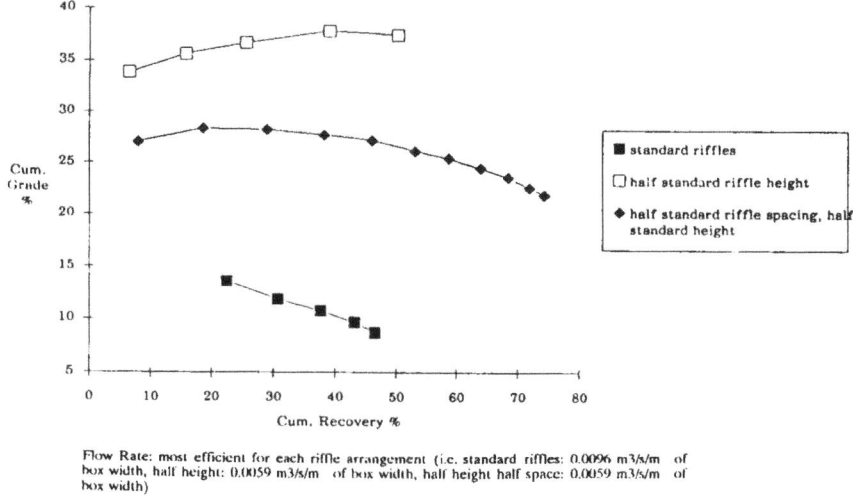

Flow Rate: most efficient for each riffle arrangement (i.e. standard riffles: 0.0096 m3/s/m of box width, half height: 0.0059 m3/s/m of box width, half height half space: 0.0059 m3/s/m of box width)

Standard riffle arrangement: Riffles 28mm high, 197mm apart

**Figure 10. Effect of riffle height and spacing on recovery and grade of heavy mineral (75 micron iron, 2000 micron gravel)**

The first was to introduce a small water flow into this dead zone area with the objective of producing some agitation in the bed of gravel and also aiding in the removal of fine gangue. This approach slightly increased the grade for coarser heavy mineral samples but with fine heavy mineral material the increased flow through the gravel bed had the effect of removing fine heavy mineral, thus decreasing both grade and recovery.

The next approach investigated was to intensify the eddy formed downstream of the riffle by directing the flow down into this area with an angled flap above the riffle. The effect of this approach is threefold: Firstly, it increases the effective height of the riffle thus improving the ability of the sluice box to initially trap material. Secondly, it stabilises the eddy downstream of the riffle and hence the gravel bed in this region. Thirdly, it induces a level of agitation in this gravel which aids both the removal of fine gangue particles and the passage of heavy mineral particles into this bed. This agitation is graded from fairly vigorous at the top of the gravel bed to very little movement at the base. This means that the passage of fine heavy mineral down through the gravel bed is aided but the quiescent zone at the base means that there is little danger of the heavy mineral being swept out of the bed. The flow patterns produced with this modification and the intensified eddies downstream of the riffles are shown in Figure 1(b).

Sample results are shown in Figure 11 (for 350 micron heavy mineral) and Figure 12 (for 75 micron heavy mineral). These figures show results for a variety of configurations. It can be seen that in all cases both grade

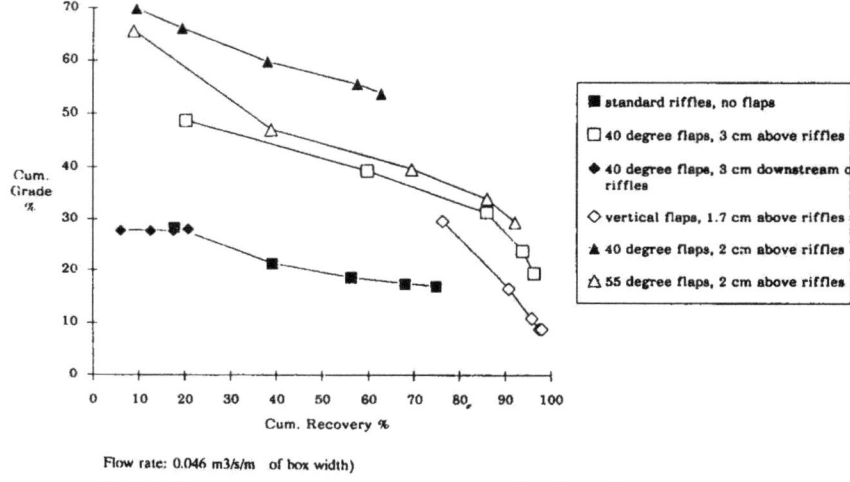

Figure 11. Effect of riffle modifications on recovery and grade of heavy mineral (350 micron copper, 2000 micron gravel)

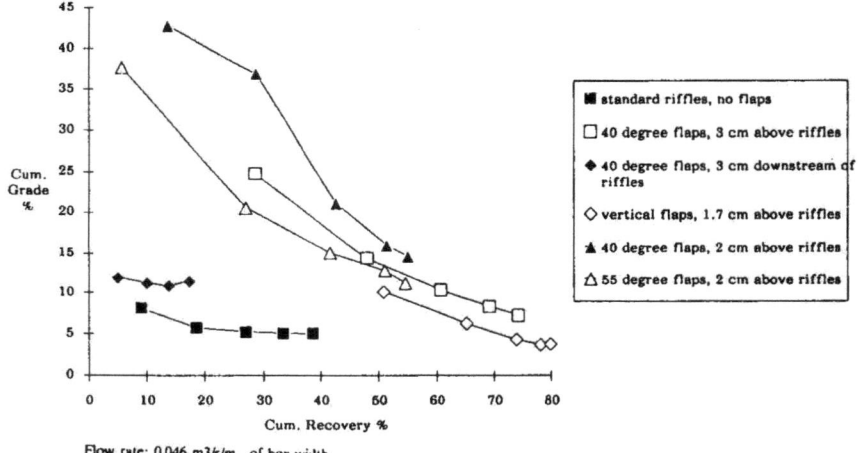

Figure 12. Effect of riffle modifications on recovery and grade of heavy mineral (75 micron lead, 2000 micron gravel)

and recovery can be increased by the use of flaps above the riffles and furthermore it is possible to choose a configuration which will maximise either grade or recovery or choose an acceptable compromise between the two. In broad terms intensifying the eddy by increasing the angle between the flap and the horizontal or by reducing the gap between the flap and the riffle will increase grade but reduce recovery and the converse occurs if we change the geometry in the opposite direction. A typical sluice box system incorporating this modification is shown in Figure 1(c).

## Studies with Natural River Gravel

In order to establish the usefulness of the above results it was necessary to establish that the same mechanisms work when the heavy mineral is present in a natural gravel with a range of particle sizes and furthermore that the system will continue to trap heavy mineral when gravel is continuously loaded into the box. It was necessary to screen out material of greater size than the gap between the riffle and the flap to prevent the gap being clogged with gravel but otherwise the gravel was untreated.

The results of such studies are shown in Figures 13 and 14. Figure 13 shows the same improvements in both recovery and grade of heavy mineral achieved when flaps are added to the riffles as described above are also obtained when river gravel is the feed material. Figure 14 shows that recovery is unaffected by continuous operation of the sluice box, at least up to the point where the box is fully loaded. Some reduction in grade with continuous operation is shown in these results suggesting that, under the conditions used, over time the box was collecting more of the larger particles of gravel, as the flow was insufficient to move them through the box.

## Application of the Results to Economic Heavy Minerals

Although we can confidently extrapolate these results to economic heavy minerals, notably gold, in qualitative terms, quantitative extrapolation is by no means a trivial question. There are a number of mechanisms which could be operating; particle settling, shear forces at the interface, movement in the settled bed (Schuring, 1977) and onset of turbulent suspension. These could all play a part in the separation mechanism. The relationships governing these mechanisms are different and hence the equivalent particle diameters derived will also be different. If, as seems most likely, the extent of particle suspension and hence transportation is the most important mechanism then the Shields criterion will be dominant (Yalin, 1977; Subasinghe and Bordia, 1990) and the ratio of equivalent particle diameters for gold and copper will be in the order of 1:2.

In order to demonstrate how the results of this research could be applied to meeting the needs of small-scale artisian miners working with simple manual methods a design for simple metal sluice box is shown in Figure 15.

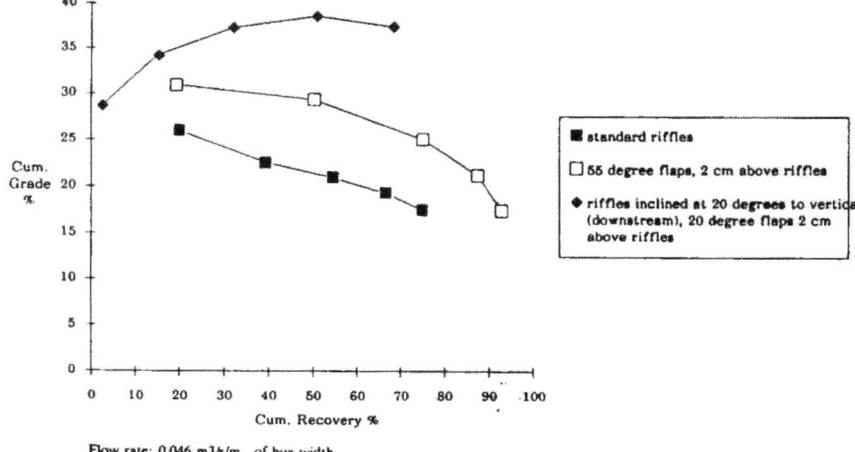

Figure 13. Effect of riffle modifications on recovery and grade of heavy mineral (350 micron copper, river gravel)

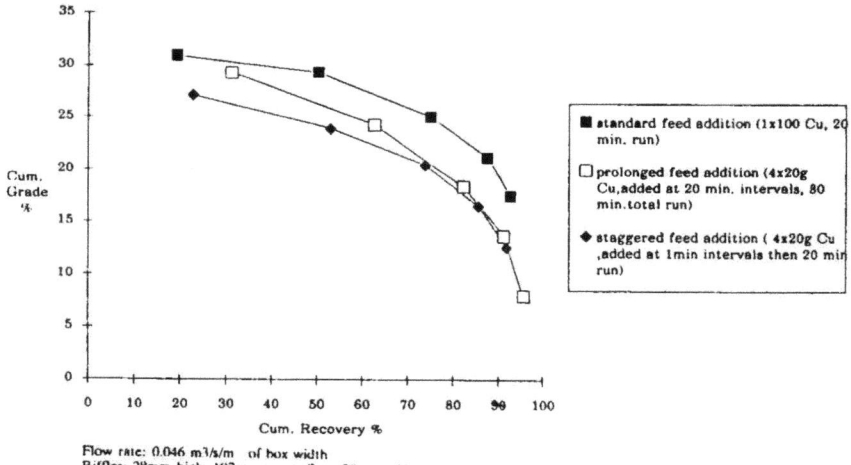

Figure 14. Effect of feed addition on recovery and grade of heavy mineral (Standard riffles, 55 degree flaps, 2 cm above riffles)

## CONCLUSIONS

1. Once the flow rate in the sluice box is great enough maintain most of the solids in suspension then solid is retained in the box in quiescent (dead) zones. The size of these zones is dependant on flow patterns within the box, which in turn are mainly dependant on box geometry. This means that, within a fairly broad range, the volume of settled

solid is independent of the flow rate and the time of operation of the box.
2. The limits of this operating range are, on the one hand when the flow rate is too low to suspend the solids, and on the other hand, when the flow rate is so great that, for that particular sluice box geometry, proper flow patterns cannot be established.
3. The gangue minerals perform an essential function in the recovery of heavy minerals in a sluice box in that the interstices between settled gangue particles provide quiescent zones where fine heavy mineral can be trapped. For this reason the sluice box is more effective in trapping fine heavy mineral than would be predicted from hydrological properties alone.
4. Because of its importance in trapping fine heavy mineral, the behaviour of the settled gangue is vital in the proper operation of the sluice box. There should be sufficient movement in this zone to allow for the release of fine gangue material and for the ingress of fine heavy mineral but too much movement will mitigate against proper shielding of the fine heavy mineral.
5. The effect of increasing box slope is similar to that of increasing flow rate except that it also causes a reduction in the dead space. This can cause a reduction in recovery.
6. A reduction in riffle height will cause a reduction in the depth of flow in the sluice box and hence the instantaneous velocity will be greater at a given volumetric flow rate. This means turbulent flow will be developed at lower flow rates. There will also be a reduction in dead space volume and an increase in the turbulence in this dead space. In broad terms, the treatment of fine feed is better conducted with low, closely spaced riffles.
7. It is possible to modify the design of sluice box riffles in the manner described in this paper so that the dead zone is stabilised and the movement in this dead zone optimised. This approach causes an increase in both grade and recovery compared to a standard sluice box.
8. The modified design can be adjusted to obtain maximum grade or recovery, or any desired balance between the two.

**ACKNOWLEDGMENT**

The author gratefully acknowledges a grant from the Australian International Development Assistance Bureau (AusAID) which made this work possible.

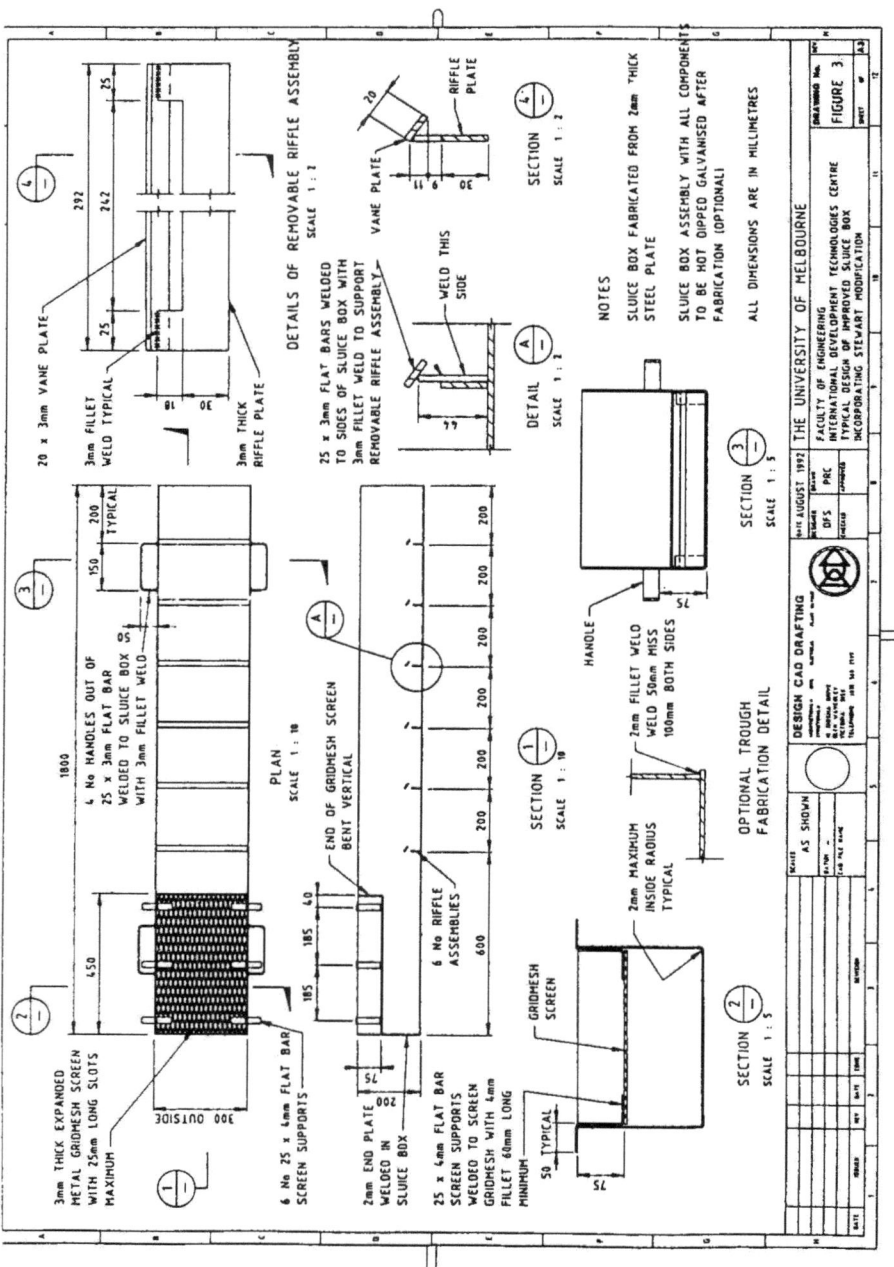

Figure 15. Design of a simple sluice box incorporating the proposed modification

## REFERENCES

Agricola, G., 1556. De Re Metallica. Book VIII, pp 322-334 (Trans. Hoover, H.C. and Hoover, L.H.) Mining Magazine: London 1912.

Burt, R.O., 1984. Gravity Concentration Technology, Elsevier, New York, 605 pp.

Fricker, A.G., 1984. Metallurgical efficiency in the recovery of alluvial gold, Bull. Proc. Australas. Inst. Min. Metall., 289: 59-67.

Kelly, E.G. and Spottiswood D.J., 1982. Introduction to Mineral Processing, John Wiley and Sons, New York, 491pp.

Meyer, F.W., 1964. Fundamentals of a potential theory of the jigging process, Proceedings 7th Inter. Miner. Proc. Cong. I, (ed. N.Arbiter) Gordon and Beam, New York, pp 75-97.

Pliny, C.P.S., c70. Natural History, Book XXXIII, p75 (English Translation Rackham, H.) Heineman, London 1938.

Schuring, D.J., 1977. Scale Models in Engineering, Pergamon Press, New York, pp 137-150.

Sivamohan, R. and Forssberg, E., 1985. Principles of Sluicing, Int. J. Min Process., 15: 157-171.

Stewart, D.F., 1990. Small-Scale Gold Mining in Developing Countries, Proceedings of the 1990 Annual Conference of the Australasian Institute of Mining and Metallurgy, Roturua, New Zealand, March 1990, pp 189-193, Australasian Institute of Mining and Metallurgy, Melbourne, Australia.

Stewart, D.F. and Gayap, J., 1989. Development Goals and Appropriate Technology in the Mineral Industry, WEEAT—World Conference on Engineering Education for Advancing Technology, Sydney, pp 130-134. The Institution of Engineers, Australia, Barton, ACT, Australia.

Stewart, D.F., 1986. Operation of the Sluice Box under Conditions of Low Water Flow, Bull. Proc. Australas. Inst. Min. Metall. 291: 81-85.

Subasinghe, G.K.N.S and Bordia, S.K., 1990. Modelling the performance of sluice-boxes, Proceedings of APCOM '90 22nd International Symposium on the Application of Computers and Operations Research in the Mineral Industry, Berlin, Republic of Germany.

Subasinghe, G.K.N.S., 1990. Sediment entrainment in sluice-boxes treating gold bearing gravel, Proceedings of CHEMICA 90, 18th Australasian Chemical Engineering Conference, Auckland, Institution of Professional Engineers, New Zealand (Chemical Engineering Group), Auckland, New Zealand.

Subasinghe, G.K.N.S., 1991. Drawbacks in sluice box operation in Papua New Guinea, Proceedings PNG Geology, Exploration and Mining Conference, Rabaul, Australasian Institute of Mining and Metallurgy, pp132-136, Australasian Institute of Mining and Metallurgy, Melbourne, Australia.

Taggart, 1945. Handbook of Mineral Dressing, John Wiley, New York. pp 11: 95-11: 104.

Wang, W. and Poling, G.W., 1983. Methods for recovering fine placer gold, Bull. Can. Inst. Min. and Mett. 70: 47-56.

Yalin, M.S., 1977. Mechanics of Sediment Transport, Pergamon Press Ltd., Oxford, pp 80-111.

# EQUIPMENT SELECTION FOR SMALL TO MEDIUM SCALE MINES

*Prabir Paul, G.C. Mishra and D.K. Panda*
National Council for Cement and Building Materials, New Delhi

## INTRODUCTION

It is a well-known myth that small scale and medium scale mines and quarries contribute major ore production all over the world. The small scale mining sector has its origin since the evolution of the mankind and is considered to be one of the largest employment sectors the world over. The far-reaching consequences of this sector over the economy of several nations, environmental set-up as well as socio-economic development had attracted the attention of even UN bodies and governments alike.

In order to implement the idea of exploiting a mineral deposit for the greatest benefit of the mine operator and the population, we must know how to utilise available ore/mineral resources. There exists a close link between the resources available in a country and the benefits which can be derived from them. "Economics" is nothing more than the conversion of one into the other. There are various resources available such as ore deposit, technology, power, water, infrastructure, management skills, workmen's skills, government aid, foreign aid and finance. A number of benefits accrue from the operation of these resources itself such as increased standard of living, capability to come into operation quickly, shorter path from mining to industry, rapid conversion of resources into benefits, amenability of certain mineral deposits to small scale mining only and at times even discovery of large orebody from starting of a small mine.

For optimum utilisation of resources, technology must be correct and rationally suitable for the circumstances. It must allow machinery to work at maximum efficiency and require minimum maintenance. It must also be environmentally acceptable. Sometimes, economic constraints often make it necessary to use an ingenious combination of second hand equipment, for example, a winch driven by an old tractor engine. In this paper, we shall mainly deal with the cement and building materials industry, especially the equipment selection criteria, the factors that govern the type, size and capacity of heavy earth-moving equipment used in the small to medium scale mines, the various types of mining equipment used and the future trends.

One important aspect to deal with is the manner of defining the order of scale of mining. Some countries use number of personnels/workmen employed per mine whereas others use quantum of HP per worker employed in the mine if it is moderately mechanised. In our country, we use production rate as the main criteria because in India the mines in the small scale industry are labour-intensive since labour is cheaper than in other countries. Thus a limestone mine, for example, in small mini-cement plant sector is required to produce around 100–300 tonnes of limestone per day for 50–200 tpd capacity plants which is considered as small scale mining, whereas for the major cement plants of medium size capacities, the limestone mines would produce around 2500–3000 tpd of limestone which is considered as of medium size. Similar is the situation in the building materials industry for construction purpose where the demand is mainly for stone aggregates, quartzites, moorum or clay for road construction etc. Other industries which operate small scale to medium scale mines are glass, ceramic and chemical plants.

## EQUIPMENT SELECTION CRITERIA

The major factors that affect the selection of mining equipment in mines/quarries are specifically noted as given in Figure 1.

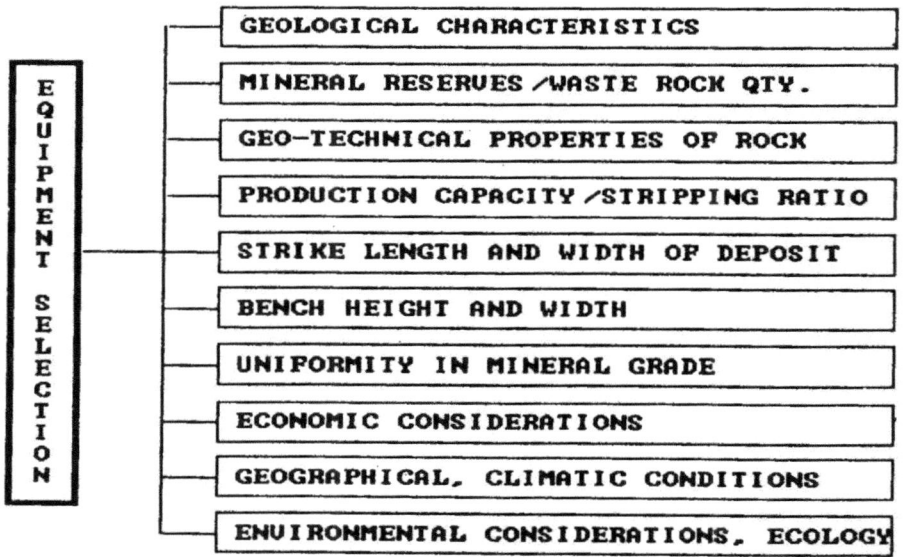

Figure 1: Equipment Selection Criteria

## SMALL SCALE MINES

For small scale mines in the range of 100–300 tonnes limestone or other building materials (stone aggregates, quartzites), the major factors that affect the selection of equipment are the production capacity, geotechnical properties (hardness of rock), life of reserves, bench height limited to mineral bed thickness and capital available.

### Mining Machinery Used in Small Scale Mines/Quarries

In small scale limestone mines, the method of working is very simple. Drilling is carried out by 50–60 mm diameter down-the-hole drill machines of reputed make and model or by jackhammers. Blasting is then carried out with small diameter slurry or NG explosives. Breaking, sizing and loading of blasted rocks are carried out manually. The sized material is loaded into small tipper-trucks of 10 tonne capacity, either manually or sometimes by small wheel loaders of 1–1.5 $m^3$ capacity. The ROM limestone is then crushed in primary jaw crushers followed by secondary hammer mill crushing to below 5 mm size.

For cleaning of bench floors and haul road maintenance or for pushing the dumped waste rock material in the waste dump yards if the deposit has overburden or side burden on the hanging wall side, a crawler bulldozer of around 150–200 HP is provided. If the deposit is of softer nature and flat, sometimes rubber-tyred dozers are provided which are more mobile and flexible. Generally, crawler dozers are preferred because of their versatility to work in any working conditions, be it flat-bedded deposit or steep gradients, whether hard rock with complex geological structure and heavy outcrop undulations on ground surface or flat land surfaces. Sometimes, if the haul distance to the cement plant or to any user plant is quite large, small width belt conveyers are used to transport the sized material after they are loaded in a moving rail-mounted hopper. These hoppers can be moved and stationed at places where the sized material can be conventionally loaded.

In the building materials industry, due to the hardness of the aggregate stones and quartzites, the breaking of the blasted rock by manual labour to smaller sizes of 6 mm, 9 mm, 12 mm, and 25 mm which are demanded as per customer's requirements takes a longer time as a result of which the productivity (OMS) goes down substantially. Thus instead of breaking and sizing by manual labour, small capacity mechanical crushers are used to crush the blasted rocks and stone and they are then subjected to screening at various fractions like 6 mm, 9 mm, 12 mm, and 25 mm as per the customers' demand. The screens are of vibrating type and because of their small mesh openings sometimes water jets are injected to clear away the clayey material coming along with the ROM material.

These crushers can be either fixed or portable (mobile) type units with the mobile ones mounted on wheels. Generally, these crushers can crush upto 30–40 tph, working 10 hrs/day. Such equipment as described above could generally be used for the small scale mines, especially in the mini-cement plant sector and also in small stone aggregate quarries. Table 1 lists out the major mining equipment used for the small scale mines. The Mine Production Cost Vs. Scale of Operation for small scale mining operations analysis can be obtained from Figure 2.

## MEDIUM SCALE MINES

In the cement industry, the medium scale mines of 2500–3000 tpd of limestone capacity are existing for a long time, though now-a-days it has been economically found viable to open up 1 million tonne or higher capacity cement plants which require around 6000 tpd limestone that come under the category of large scale mechanised mining operations, provided large capital funds are available. However, in this paper, we will be dealing only with the medium scale mines of 2500–3000 tpd which have been in mining operations for quite sometimes.

Table 1: List of Main Mining and Crushing Equipment for Small Scale Mines in Mini-Cement Plants and Building Material Industry.

| Sl No. | Item | Capacity | Remarks |
|---|---|---|---|
| 1 | DTH drill machine (wagon drill) | 50–60 mm dia hole | —— |
| 2 | Wheel loader | 1–1.5 m$^3$ | For mine capacity more than 200 tpd limestone |
| 3 | Tipper-truck | 10-T | —— |
| 4 | Bulldozer (rubber-tyred or crawler-mounted) | 150–200 HP | —— |
| 5 | Jackhammer | Medium weight | —— |
| 6 | Portable air compressor | 250–300 cfm at 100 psig | —— |
| 7 | Crusher (jaw crusher and hammer mill-2 stage) | 40–50 tph | For crushing to sizes below 5 mm in mini-cement plants. |
| 8 | Crusher (i) jaw crusher (ii) cone crusher | 30–40 tph 15–20 tph | For crushing to sizes below 25 mm in building material industry. |
| 9 | Vibrating screen-Double/triple deck | Varying capacity | For screening to sizes of 6 mm, 9 mm, 12 mm, and 25 mm in building material industry. |

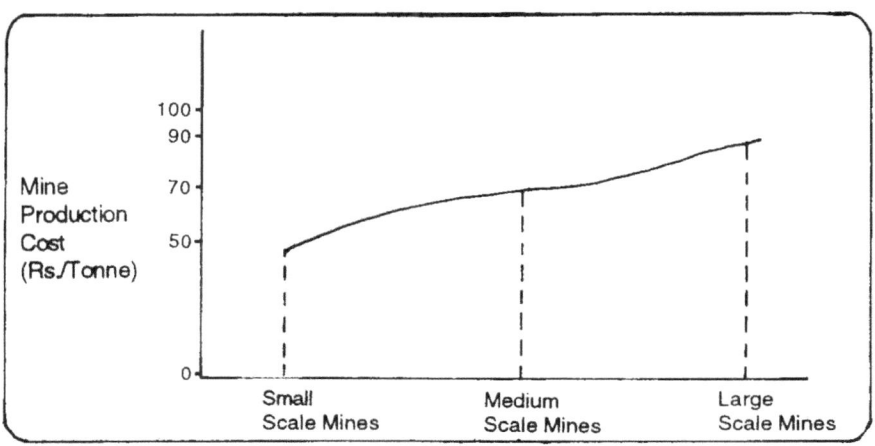

Figure 2: Mine Production Cost vs Scale of Operation

The quantum of mechanisation in medium scale limestone mines is controlled, yet highly productive. Earlier cable/rope shovels were used for excavating the blasted rock but now-a-days they are being gradually replaced by small to medium size hydraulic excavators whose operating and maintenance cost is much lower and whose productivity is much higher as compared to the cable/rope shovels of yesteryears.

## Mining Equipment Used in Medium Size Mines/Quarries

The drilling operations are carried out by medium size down-the-hole (rotary-percussive) wagon drills of 112 mm dia., mounted on tyred wheels. The drill machines which are less capital-intensive have proved to be very productive for the medium scale limestone mines and are also quite flexible in operation. For blasting, the slurry or emulsion explosives are used or the more economical ANFO explosives primed with boosters are used if blastholes are not watery. Sequential blasting which has become very popular now-a-days is also carried out if the number of holes to be blasted at time is large or where the limestone and waste rock lying adjacent to each other are to be separated out and segregated in 2 heaps at the mine face so that excavation of the blasted muckpiles does not cause any dilution or contamination. In other words, rock segregation is possible by rational utilisation of sequential blasting technique through the use of the sequential blasting machine which besides producing better fragmentation also reduces environmental hazards like ground vibration, fly rock, backbreak etc.

Bench heights are generally kept at around 6–7 metres generally for the medium size mines so that small to medium size hydraulic excavators

Table 2: List of Major Mining and Crushing Equipment for Medium Scale Mines.

| Sl. No. | Item | Capacity | Number of units | Remarks |
|---|---|---|---|---|
| 1. | Hydraulic Excavator | 2.5 m$^3$ | 3 | —— |
| 2. | Rear dump-truck | 25-T | 12 | —— |
| 3. | Tyre-mounted wagon DTH drill machine | 112 mm dia. hole | 4 | If deposit is simple, tyre-mounted wagon drills (112 mm dia.) may be used. |
| 4. | Portable air compressor | 400 cfm at 150 psig | 4 | —— |
| 5. | Bulldozer | 250–300 HP | 1 | —— |
| 6. | Wheel loader (FEL) | 1.5–2.0 m$^3$ | 1 | —— |
| 7. | Rock Breaker (mounted on small hydraulic backhoe excavator) | 20–30 boulders per hour | 1 | —— |
| 8. | Crusher (rotor impact breaker) | 300 tph | 1 | For crushing to sizes below 75mm or 25 mm in cement sector. |
| 9. | Crusher (i) jaw crusher | 200–300 tph | 1 | For crushing to sizes below 25mm in building material industry. |
|  | (ii) cone crusher | 100–120 tph | 2 |  |
| 10. | Vibrating Screen Double/Triple deck | varying capacities | 1 | For screening to sizes of 6 mm, 9 mm, 12 mm, 25 mm in building material industry. |

of 2.0–2.5 m$^3$ capacities can be used to match with their maximum cutting height range. Medium size dump-trucks of 20–25 tonne capacities are preferred to match with the above excavator bucket capacities. Moreover, these dump-trucks have higher speeds and are quite flexible and the spare parts availability is easier.

The dump-trucks of above capacities are not very capital-intensive and their operating and maintenance expenses are quite low. Caution should, however, be implied for these haul units by not overloading them excessively or loading oversize boulders because these dumpers are fitted with ordinary mechanical leaf spring suspension system whereas the bigger capacity dump-truck have hydro-pneumatic suspension which can absorb higher payloads on rough haul roads. Bulldozers of around 250–300 HP are preferred to the smaller ones because of heavy workload in the waste dumpyards.

In the processing and sizing of limestone, medium size crushers of 300–350 tph are being used for medium scale mines. The most common crusher used in the limestone mines for the cement industry is the rotor impact crushers because in this sector all ROM material has to be crushed either below 75 mm or below 25 mm, depending upon the hardness of the rock. As limestones are associated with clays and free silica (quartz), in most of the plants a grizzly with a scalping screen 25/10 mm mesh size beneath is used to scalp out the −25 mm or in some cases −10mm size rejects/off-grade deleterious materials from the ROM limestone before the large size limestones of better grade are crushed for cement manufacture.

In the building materials industry, the medium scale mines of stone aggregate, quartzites etc. are being exploited for the construction of large civil works like dams etc. and industrial buildings, though their numbers are fewer as compared to the limestone mines of the cement industry. For irrigation works also the rock excavation is quite large. The mining machinery used in these mines are similar to those of the cement sector but the processing and sizing equipment differ. In this sector, jaw crushers of 200–300 tpd together with small/medium cone crushers are preferred, since the desired product sizes of stones are more. A double/triple deck vibrating screen will be used to produce a number of smaller size stones, namely 6 mm, 9 mm, 12 mm, 25 mm etc. as per the customer's choice.

Such equipment as described above which are being used in the cement and building material industries are being recommended for use in the medium scale mines and quarries of this sector. Table 2 gives a list of major mining equipment required for a limestone mine producing around 2500 tpd of limestone and another 500 tpd of waste rock overburden.

## PRODUCTIVITY IMPROVEMENT IN MEDIUM SCALE MINES

The productivity of mining equipment in the cement sector can be increased by set standard norms of availability and utilisation which were suggested by NCB, New Delhi and to which positive and encouraging response was discernible from some of the cement plants. Some of these norms for major mining equipment are given in Table 3 as a guideline for achieving higher productivity.

The Mine Production Cost vs. Scale of Mining Operation analysis (Figure 3) shows the mine production cost for different systems of mining operations i.e. fully-mechanised, semi-mechanised and labour oriented. Figure 4 illustrates a histogram of availability/utilisation of mining equipment.

## INPIT CRUSHING—CONVEYING TECHNOLOGY

Due to the increased fuel oil prices, the usage of dump-trucks for long haul routes of 2–3 kms is becoming increasingly expensive, thereby increasing the mine operating cost considerably. For last 2 decades or so, mine operators and owners of the developed countries of the world started

Table 3: Availability and Utilisation of Mining Equipment

| Sl. No. | Equipment | Capacity | Availability (%) | Overall (Net) Utilisation % |
|---|---|---|---|---|
| 1 | Hydraulic Excavator | 2.0–2.5 m$^3$ | 75–80 | 65 |
| 2 | Rear Dump-truck | 15–25 T | 70 | 65 |
| 3 | Wheel loader | 1–2 m$^3$ | 70–75 | 60 |
| 4 | Bulldozer | 200–300 HP | 70 | 60 |
| 5 | DTH wagon drill machine | 112 mm dia.hole | 70 | 65 |
| 6 | Crawler mounted DTH drill machine | 112 mm dia. hole | 75 | 65 |
| 7 | Portable air compressor | 200–300 cfm at 150 psig | 70 | 65 |

using the mobile/semi-mobile crushers of the inpit crushing system in mine production planning. These inpit crushers were used in conjunction with belt conveyers so as to reduce or eliminate the dump-truck transport from the quarry to the conventional crushing plant located further away from the mine faces.

In our country, a slight modified version of inpit crushing by permanently erecting/constructing the crushing plant at the furthermost end of the pit boundary limits on a non-mineralised zone has been in operation for the last 3–4 years. From this permanently located plant, a belt conveyer is laid out to carry the sized/crushed material to the user plant. Whereas mobile crushers mounted on wheels/crawlers can be used for long mining faces at a stretch, semi-mobile crusher module mounted on wheels can be used for inpit crushing of medium size mines if a number of working benches or mine faces are in operation. In such a case the crushers can be stationed at a safe place inside the pit after jacking them at a particular safe spot for at least 4–5 years, one bench level below the mine working areas as a thumb rule so that the dump-trucks from various mine faces will carry material loads downwards in favour of gradient and then unload the material into the hoppers of these semi-mobile crushers, thereby reducing the haul distance and consequently the fuel consumption. The crushed material can then be transported by mobile transfer conveyers and shiftable conveyers up along the pitwall and onto an overland belt conveyers right upto the user plant. Some cement manufacturers or mine owners prefer to permanently install and erect an inpit crusher at the further end of the pit on a non-mineralised zone and very near the plant as explained earlier, thereby minimising fuel consumption and hence the haulage cost.

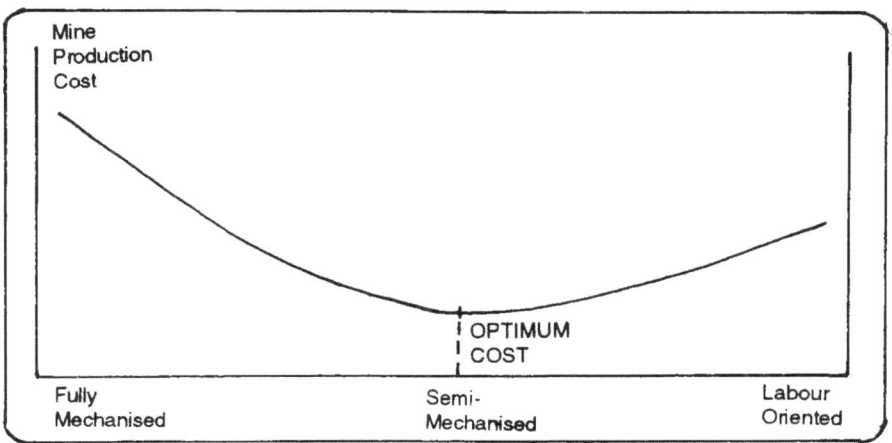

Figure 3: Mine Production vs Scale of Operation for Small Scale Mining Operation

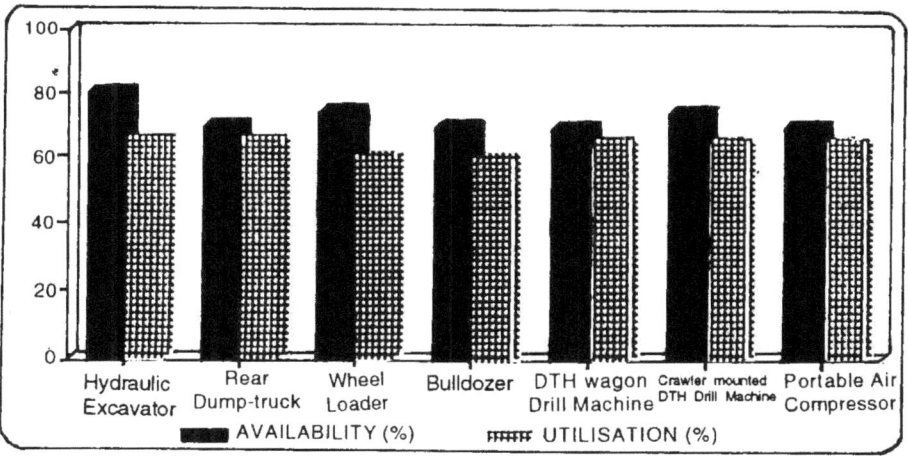

Figure 4: Availability and Utilisation of Mining Equipment

## APPLICATION OF CONTINUOUS SURFACE MINING TECHNOLOGY

The concept of using continuous surface miners (CSM) for medium scale limestone mines in the cement industry is not far off. Continuous surface miners are basically highly productive, economically operative, environment-friendly and quality control mining equipment comprising large cutting or milling drum in the front with cutting picks (tools) fixed all around drum circumference. These machines cut and load the insitu rock directly. Behind the drum is the loading and discharge belt conveyers. The whole unit is mounted on 3 or 4 crawler mechanisms.

Advantages of these surface miners are multifarious including 30–40% saving in operating costs and ability to cut strips of ore or mineral upto 25 mm thick without causing any contamination or dilution with the contact rock — a great advantage over the conventional excavating equipment in quality control, especially in non-uniform graded heterogeneous deposits.

In our country, smaller models of these equipment are being used by the cement industry for the limestone mines with capacities of 120–150 tph, having softer rocks (less than 50 MPa compressive strength). Since there is no drilling and blasting, the environmental hazards of gaseous fumes, dust, ground vibration, fly rock etc. are all eliminated. Their only restriction is the high prohitive capital cost, but that is offset by the large saving in operating costs, thereby reducing the payback period of the equipment over its economic life of around 15 years. In some countries of the world, a large cluster or group of medium scale and large mines are purchasing these equipment to use them on group cooperative system of owning so that all the mines get their chance to utilise the CSM's profitably and thus share the large capital investment on a proportionate basis.

## CONCLUSION

Whereas large scale mines are quite capital-intensive and also require large mineral resources to start with mining operations, the small and medium scale mines by selecting proper mining equipment rationally can be a major supplier of minerals/ores for several industries which may not require large mineral/ore reserves. A close look into the geology and other geo-technical or mining parameters of a particular deposit and its suitable use in the small scale/medium scale industries is very essential to decide the mineability of such deposits.

## ACKNOWLEDGEMENT

The authors wish to thank Dr. S.P. Ghosh, Director General, National Council for Cement and Building Materials for giving permission to publish this paper. Mr. Parveen Kumar is also highly acknowledged for his support in preparing this paper.

### REFERENCES

1. P. Paul, G.C. Mishra and D.K. Panda (1995). "Computerised Mine Planning for Highly Mechanised Quarry Operations". 14th AFCM Technical Symposium—Modernisation in Cement Industry. Kualalampur, Malayasia.
2. Raj K. Singhal (1988)." Mine Planning and Equipment Selection". AA Balkema/Rotterdam, 1988.

3. G.C. Mishra, D.K. Panda and P. Paul (1995). "Computer-aided Deposit Evaluation and Mine Planning—A Case Study". Proceedings of National Seminar on Status of Mineral Exploration in India (MINEX-95, Delhi Chapter), New Delhi. pp 138-147.
4. A.G. Pasamehmetoglu, S. Eskikaya, C. Karpur & T. Hizal (1994). "Mine Planning and Equipment Selection" A.A. Balkema/Rotterdam.

# IDENTIFICATION OF TECHNOLOGICAL PROBLEMS IN SMALL SCALE MINING IN TANZANIA; A MOVE TOWARDS POVERTY ALLEVIATION

W. Mutagwaba, A. Mlaki and R. Mwaipopo-Ako
State Mining Corporation, Dares Salaam, Tanzania

## INTRODUCTION

Small scale mining in Tanzania, involving individuals, families and ad hoc groups has existed for a long time now. Its formal recognition is however recent following the recognition of its significant contribution to the national economy. 76% of all mineral exports earnings in 1992 were from gold mining by small scale miners. Also, gold bought from small scale mining between 1990 and 1994 was 13.84 tons. Taking into consideration the social and economic levels of the country, this sector could be seen as a response by people to handle poverty themselves in its various forms. This is particularly true taking into consideration the decline in traditional sources of livelihood, e.g. in agricultural production, fishing etc. Nevertheless, its 'informal' operations has allowed it to attract many people because of the relative easiness of entry, luring people by visions of attractive gains. Thus it is also important in offering productive employment.

Among its salient features include, minimal investment in training and skills, it is labour intensive, and attracts cheap labour, therefore gives access to employment to many people, it is basically unregulated, i.e. it operates outside the government structured administrative and institutional framework therefore one enters as opportunities arise, and may evade income tax.

Small scale mining in Tanzania is however, characterised by low efficiency, low production, low wages poor working conditions and others (Mutagwaba et al. 1996). According to Mchaina et al. (1987), this is because the sector suffers from lack of modern technology, financial support, technical assistance, training and education, and, accessible sources of information to assist in its development. Ngonyani (1996), associates the dismal performance of small-scale miners to: (a) lack of permanent indigenous mining community; (b) lack of investment capital; (c) restriction of small-scale to Tanzanian citizens only; (d) lack of attractive mineral resources base; and (e) lack of technical expertise.

However, the inadequacies in small scale mining may be far from being fully realised not only because of the many inadequacies in the process itself, but also from the regulations and practises that attempt to manage it. For example, the lack of a legal and fiscal framework, and because of "inefficient production, processing and marketing arrangements" (Jennings, 1993:1). Hence its analysis involves a "much more complex phenomenon with economic, cultural and socio-political dimensions" (Wagao and Kigoda, 1993:4 as cited in Cooksey, 1995:74).

Technological considerations, however, feature as very significant because of their direct implication on output levels, safety and environmental aspects, which in turn determine the continued well-being of the populations involved in artisanal mining. In addition, they are not only geared to improve working conditions and output but also focus on the quality of work rather than just the quantity. Any increase in the quality of human effort put to artisanal/small scale mining will be more than matched by the quantity, quality and value of output due to more efficient production and higher value added at the primary processing stage prior to sale (Jennings, 1993:30).

This paper is based on a survey conducted in 5 mining areas in Tanzania, involved in gold, Tanzanite, ruby and diamond mining. The major focus was to explore and identify technological aspects related to artisanal/small scale mining in Tanzania. In addition, the paper also highlights some institutional and organisational considerations affecting this sector and their inter-relationship with technological constraints. Five sites were visited; two gold mining areas, namely, Mugusu and Rwamagaza both located in Geita district, Mwanza region, one diamond mining site, namely Mabuki in Misungwi district, Mwanza Region, and two gemstone mining areas. These are, Merelani in Simanjiro District, Arusha Region and Ng'ongolo in Matombo, Morogoro Region. Whilst the first is known for mining of Tanzanite, Ng'ongolo is famous for its rubies.

## Definitional Problems

Several definitional attempts on what is or what constitutes a small scale mine have been made. Some of these are based on: investment costs, labour requirements, ore production rates, size of concessions, amount of reserves, annual sales, or any combination of the above. The World Bank in its document "Tanzania Mining Sector Review" acknowledges the definitional problems that arises from the use of the term "small-scale" mining (WB, 1990). It categorised their activities as varying what might be termed "subsistence mining" performed by an individual or family recovering minerals to support their basic rural life-style-to small-scale, essentially, non-mechanized mining. The report however, continues to use the term "small-scale in what it terms "the Tanzanian Context" to refer to groups of persons, be they, cooperatives or some other type of enterprises, mining by simple methods in a manner largely uncontrolled by the authorities.

In the Tanzanian context, the definition of small-scale mining can be found in a policy document dated September, 1983, and entitled "Small-scale Mining Policy". It defines small-scale mining as having the following characteristics:

- Labour intensive with low initial capital (below Tshs. 500,000/=)
- Do not require skilled labour and/or specialised technology.
- The gestation period from exploration to production is short.

The setting of limits of production has often raised objections because of its inappropriateness as a common measure for all minerals. Likewise, is the unreliability of setting of a minimum base as cost for investment.

The Southern African Development Coordination Unit (SADC) Mining Coordinating Unit subdivides small-scale mining into 3 categories namely:

(a) *Micro-scale mining*: Manual mining with simple tools without using mechanical energy;

(b) *Manual mining*: well organised using some mechanical energy and plant. Required investment are between US$ 10,000 and US$ 100,000;

(c) *Industrial small scale mining*: small mines using modern, adapted technology. Required investment are US$ 200,000–3,000,000 or more.

It is beyond the scope of this presentation to delve deeper onto such discussion. However for purposes of situating the discussion within the context of technological aspects, it is suffice to mention a few commonly identified features; the absence or low degree of mechanization due to high proportion of heavy manual labour, low safety standards, -poorly trained personnel, -lack of technical personnel in the plant, resulting in deficient planning in both mining and processing activities, -comparatively poor utilization of resources due to non-selective mining of high grade ores and poor recovery, -low pay scale, low work productivity, -periods of non-continuous mining, as a result of mining seasonally or when world market prices reach a certain minimum level, chronic lack of capital, -some illegal operations due to mining without concession rights (Priester et al., 1993).

For purposes of practical relevance, this paper refers to the micro-scale or subsistence mining sub-sector. The term "artisanal mining" is sometimes also adopted to refer to those activities carried out by individuals, families, and/or adhoc groups (some forms of cooperatives) of local peasants the majority of which have no technical know-how or appropriate tools.

## WORKING ENVIRONMENT

Newman (1987) highlights that for such small workings to be profitable two basic conditions must exist. "Either the mineral price should be high relative to the operating costs, or the mineral grade high", and the reliability of ore reserves (Newman, 1993). These aspects may cater for price fluctuations, taxation and poor production methods. But due to the unpredictability of this venture politically and financially, and the limited or unbalanced information these miners may have on ore reserves (because they seldom embark on initial exploration), too much economic risk is involved when they embark on production or exploitation. In any case, there is a high correlation between the risk involved and level of technology applied (in addition to the organisational and institutional deficiencies).

Being a largely labour intensive operation, it has a greater impact on **employment** than the large scale mechanized mines. Lwakatare (1993) estimated that 'artisanal' mining involves 280,000 or more miners on fulltime or part time basis, and Chachage (1995) notes that "at a very conservative estimate of 10,000 per mining site, it is possible that there were (by 1993), about 900,000 people involved in small scale mining and auxiliary activities. A recent study conducted by TANDISCOVERY Ltd. contends that more than 555,000 people are directly involved in mining activities throughout the country (TANDISCOVERY, 1996). While the figures of the population actively involved in small scale (artisanal) mining can be grossly exaggerated without a properly conducted census, it can be clearly noted, that it involves a significantly large number of people.

Artisanal/small scale mining has also been a significant source of livelihood for most of the mining population. Although it may not be realistic to establish a fixed amount of income for such activity, estimates on incomes indicate that at an average, these incomes are more than the current government minimum salary of Tshs 30,000/= per month (about $50). TANDISCOVERY (1996) estimated monthly incomes as follows, the claim owner in reef gold mining areas gets about $730.2, pit owner $420.0, mine workers groups (5 people) get $855.8 or $171.76 per individual miner. Another study conducted by the authors, estimated Tshs. 200,000/= per month (about $333.4) to be the income of the Matombo ruby miners per pit. This amount when divided by an average of 6 people per pit was Tshs 30–40,000/=. But given the fact that miners occasionally scoop stones worth Tshs 1–3 million, their overall income is a lot higher.

The existing **organisational and Institutional framework** governing artisanal/small scale mining operations in the country include, the Ministry of Energy and Minerals (MEM) which is the government organ for administering the mineral sector through the Mineral Resources Department (MRD). The MRD is responsible for overall development, promotion and coordination of mining activities in the country. In addition, there are 18 Zonal offices and a number of regional and district offices which serve as implementing structures of the MRD. Basically, the MEM and MRD are adequate for the formulation of policy guiding mineral exploitation in the country. However, they are constrained by a number of factors including, shortage of resources, personnel, a small budget, lack of operational equipment (vehicles technology).

**Cooperative formation** was initiated with the envisaged benefits of improving individual mineral exploitation. Hence, Miners associations and groups have been established in most areas with active small scale mining. Their roles and functions include, mobilisation of the artisanal / small scale miners, representing them in mining disputes, providing assistance or organising marketing, transportation etc. These associations to some extent have been commended by their role in encouraging the miners to apply for formal registration and peg claims, and in some cases they indirectly assist the Zonal offices in overseeing the smooth functioning of activities. Of late however, they have been plagued by problems relating to weak leadrship, declining commitment by members, and shotage of funds.

**Financial support in credit:** There are several institutions and organisations with the capacity to provide loans or credit support to this sector. Among the NGOs identified are, Poverty Africa, The Business Center, Pride Africa and World Vision. Some institutions formerly identified as willing to provide credit to such a sector include the National Bank of Commerce the Tanzania Investment Bank (TIB). However, most of their demands in terms of collateral and security could not be met by a majority of the miners. The Postal bank is a recent possibility which has not yet been explored.

## Technological Cosiderations

Technically, mining projects go through certain stages before commencing production. These usually involve reconnaissance, prospecting or looking for minerals, evaluation which involves collection of samples and testing for particular minerals which if found, feasibility studies are then conducted to establish exploitation and beneficiation methods under existing geological conditions and hence establish viability of the project. These processes are followed by mine development and construction of the processing plant and other facilities before commencement of production. Artisanal/small-scale miners in general neither have the financial resources nor the technical know-how required to pursue the above activities. However, these miners have a knack of scouting the wilderness in search of minerals hence their reference as barefooted prospectors.

The remoteness of most of these areas have increased the significance of these prospectors as they provide basic geological information to geological survey teams. Consequently, their uncoordinated barefooted exploration provide fiscal benefits to the government from remote reserves that would otherwise be classfied as uneconomical. However, in most cases artisanal miners do not work on new discoveries. This is very common here in Tanzania where most artisanal workings are found in areas rejected by exploration teams as being uneconomical, e.g. pillars in old closed down mines, piles of tailings from old mines, and other such areas. For example, according to Hollaway (1993) nearly a quarter of all small-scale mine workings are developments on an old formal mine. Mine workings at Siroti Simba, Mobrama (Nyabizama) and Kibaga provide good examples.

New discoveries, although rare, are sometimes made by pure luck. Some potential reserves have been stumbled by farmers, herdsmen or people digging latrines and encountering small nuggets in the process. Examples of these include discoveries at Bulyanhulu, Nyambegana and Serengeti in that sequence. The Bulyanhulu deposit which was first stumbled upon by artisanal miners in 1976 has turned out to have reserves of up to 4.3 million tonnes grading 10.76g/t of gold, 12.05g/t silver and 0.66% copper, were established (Mutagwaba, 1993).

The production process of these artisanal/small scale miners involves both prospecting, evaluation and production, undergone concurrently. Pitting and trenching is usually production development. The pits and trenches are usually abandoned once reserves are thought to have been exhausted or it becomes difficult and dangerous to mine any further. The lack of technical know-how and support, is the main cause of most artisanal technological problems associated with mineral extraction and beneficiation.

## OBSERVATIONS

Following the survey on which this paper is based, the following were observed. There are a lot of similarities in techniques for mineral extraction and beneficiation used in these areas. Nevertheless, some experienced problems are unique to a particular site.

Mineral extraction is done in pits whose sizes and spacing are not based on any technical judgement but on miners' intuitive judgement. Pits are dug straight down until secondary enrichment is encountered. Miners seem to have gained experience in the local geology such that pits tend to change direction as they detect loss of mineralisation. Once mineralisation has been encountered it is pursued along the strike through crosscuts and roadways. The width of the roadway are in most cases dependent on the width of the orebody. After mineralisation has been encountered the pits tend to change direction along the dip of the orebody.

As such, all development is restricted within the mineralised zone. The method of extracting the mineral bearing rock is by manual means. By using hammers, chisels and single-sided picks also known as sokomoko, the ore is slowly chipped off before being hoisted to surface. Productivity tends to be reduced substantially as mining goes deeper into primary mineral enrichment which is in harder rocks. Sometimes explosives or application of rock drilling machines are used.

Equipment used for mineral extraction is very rudimentary, most of which are fabricated at the mining site. At the start of pit excavation i.e excavation in soft weathered ground, the following equipment is used: (i) Shovels, (ii) Picks, (iii) Forged old drill steels with one end sharpened. As pits get deeper, (iv) Ropes, (v) Sacks are used. As more hard rock is encountered, (vi) hammers, (vii) Short steels with one end bevelled (also known as "Ponch") or chissel, (viii) One sided picks (known as sokomoko in Merelani), (ix) Buckets.

Blasting operations are sometimes contracted out to someone with a blasting certificate. The contractor usually provides the explosives, detonators, stamping material, etc. 2–4 feet (0.6–1.2 m) length holes are usually drilled. However, many miners still break the rock by manual methods. Pits are usually excavated with notches on either side to facilitate movement of men between surface and underground.

Extraction in diamond mining is done by pitting in shallow wells. The gravel is thrown to the surface with a shovel. No hoisting or tramming is required. Excavated gravel is washed during the rainy season when pits are filled with water. Tanzanite extraction is also done through pitting mainly on hard rock. Some mechanisation has easened the process of extraction. Compressors, drilling machines and the use of explosives is extensive.

Water problems in pits is often encountered in areas were the water table is very high. A few deep pits on site have encountered water at 100 m and below. Most pits encounter water at a depth of 12–15 m. In most areas, there were no pumps observed on site. As such, water is removed from the pits by using a bucket and rope. A bucket is manually filled with water at the bottom of the pit and then hoisted to surface for disposal. Given the topography and the fact that some pits a located in abandoned quarries, the disposed water may find its way underground again.

**Mineral Beneficiation**

In gold beneficiation, the primary ore extracted from underground has to be processed before recovery of the metallic gold. Processing of the ore goes through different stages and thus uses different equipment. For example, in Ore preparation processes, size reduction (crushing and grinding) is carried out by using hammers followed by further size reduction in a hardwood mortar and half axle pestle. Grinding stones are used for fine grinding. The raw ore is crushed manually by using hammers as an initial size reduction stage. The product size of about 1 cm is crushed further in a hardwood mortar a motor vehicle half axle as a pestle. Depending on the type of ore, sometimes fine grinding is necessary. Fine grinding to less than 100 micro-meters is carried out by using a grinding stone, a job usually carried out by women and children.

In gold production/recovery processes, in cases of high grade ores (normally determined by miners on the basis of colour), the slurry made from crushed ore is panned directly to recover the metallic gold. Low grade ores are processed by running the slurry over the inclined sacking strakes (clothed tables without riffles). Gold concentrate collects over the cloth or sack and is recovered by washing the cloth in a bucket of water. This stage is usually carried out on river banks due to high water requirements. The concentrate is then mixed with mercury to form the gold-mercury amalgam. Recovery of the metallic gold bullion is then carried out by heating the amalgam on a pan over an open fire in order to evaporate mercury.

In alluvial deposits, gold recovery is through panning and a few sacked strakes followed by mercury amalgamation. The gold-mercury amalgam is then evaporated on a pan on an open fire in order to recover the gold bullion. In one area, it was estimated that on average 12 grams of gold is recovered from 100 kg of crushed ore. This implies an ore grade of about 120 grams per tonne.

In diamond beneficiation, a simple seiving system is used to recover diamonds from the gravel and clay. This may take up to 3 months retreating gravel, which is later sorted in a pan. Tanzanite is recovered through controlled blasting and the use of hand hammers. Panning is sometimes done to recover smaaller pieces which could not be collected through sorting underground. In Ruby beneficiation, hand hammers and sieves are used to process ruby. Hand sorting is also applied.

The beneficiation equipment used in gold can be catergorised as follows: (a) Crushing and Grinding: Hard Hammers (4 kg); Hardwood Mortar; Half Axle pestle; (b) Separation (Concentrate Recovery) Steel Pans; Gold Pans, Sacked Strakes; Buckets; Amalgamation Retort (Available but no usage could be established).

The beneficiation equipment in Tanzanite production include, small hand hammers and steel pans.

## PROBLEMS RELATED TO APPLICATION OF TECHNOLOGY

The identified problems related to the application of technology include:

- Miners lack adequate knowledge of the type of rocks and geological conditions. As such, the depth, width, inclination, grade, etc. of the orebody are not known.
- Location of pits is done haphazardly without any technical consideration and thus endangering their stability. Some pits are located within the zones of influence of others. In addition, the sizes of pits are too small to warrant arrangement of a formal shaft i.e. with manway, hoisting chamber, services, etc.
- supports are often not used. Where they are used, they are inadequate or weak. In addition, the high cost of timber (logs) encourages mining without support.
- In most sites, the equipment used is inadequate for the job. Where mechanical equipment has been introduced, it is underutilised.
- In relation to Mine Ventilation, there is a complete lack of adequate ventilation which has sometimes led to fatal accidents, e.g. suffocation by CO gases. Sometimes, expensive equipment like compressors are used to ventilate deep pits e.g. Merelani. Apart from being an expensive way to ventilate a pit, a substantial amount of moisture is added into the mine air.
- Although some mechanical equipment has been introduced in some sites, the majority still hoist the broken rocks manually. As a result tramming and hoisting is one of the major delay points of the production cycle.
- There is an obvious lack of water pumping and water storage equipment.
- Heavy sedimentation and mercury pollution occurs especially around gold mining areas.
- There is inadequate knowledge on mineral beneficiation process and principles, especially on the processes of mineral dressing and metallurgical principles. This can be exemplified by some gold miners rejection of the use of simple retorts for mercury distillation after amalgamation for fear of loosing the gold in the retort.
- Miners lack the appropriate know-how to recover smaller gemstone particles. In addition, they lack techniques for processing the gemstones further after sorting as a means of increasing their values (value added measure).

The seriousness of these problems is aggravated by the fact that, there is an obvious lack of technical advice to miners on various necessary issues. For example, mine site inspections are very rare and when done they are too casual to have any advisory impact. In addition, there is a clear lack of information pertaining to geology, marketing facilities and equipment.

— Safety matters, e.g. on general health and safety, transportation and storage of explosives, mine support systems, etc, are not well addressed or exposed to the miners advantage.
— Finally, there is a complete lack of training facilities and programmes for the miners.

Moreover, various complementary factors work against the possibility of investing in technology in small scale mining. These include:

— The nature of Social Organisation and Division of Labour: The nature of the division of roles and tasks which are associated to ownership and the multiple processes involved, have created a system which is exploitative in that, the division of earnings between the claim holder/pit owner and the mine workers is often very unfair, and not commensurate to the type of the work involved. In addition, most claim holders invest nothing in the area but rip the highest benefits. The mine workers are regarded as people with a non-binding contract with the pit owner. For example, the claim holder may decide to lease a pit to another person with hard cash on short term basis with no consent of the miners. This discourages many to invest in technology.
— Poor accounting and financial management systems: despite the significant amount of money the artisanal/small scale miners may realise, many fail to invest in profitable ventures because of the urgent need of cash at that time or due to lack of banking services available or because of lack of appropriate information on how to invest.
— Vulnerability, in terms of the delicate social and political environment the miners operate within. Conflicting property (land) ownership policies render small miners vulnerable to dis-ownership when bigger or state interests overrule ownership patterns.
— Institutional and Organisational Aspects: There is still very little concerted efforts directed on alleviating the small scale miners situation that is supported by the MEM. The sequencing of steps to operationalise policy directives is still contradictory and not transparent. The mining policy (1996) is now being processed to create that enabling environment which include the promotion of the use of appropriate technology. However, there is too much centralisation of functions and responsibilities by the Ministry and its offices which may affect the possibility of facilitating this proposal to the miners. Other related problems are related to the inadequacies in the banking and credit system. Access to loans and credit is a problem.

In addition, there are many deficiencies in the marketing system which has a direct implication to the miners financial capacity. Firstly, there are limited opportunities for marketing within the production areas. In addition, since 1993, the Central Bank through the National Bank of Commerce (NBC) has stopped purchasing gold from individual miners. This situation has contributed to the expansion of 'illegal' mineral dealers (and possible loss of national revenue). Secondly, the prices offered (for example, for gold) are below world market prices which discourage the miners who resort to informal marketing channels. In February 1996, the prices offered for gold by the National Banking institutions was Tshs 4500/=, about $7.5, while the price offered at the parallel market was between Tshs 6000/= and 7000/=, about $10 and $13 (Mutagwaba et al., 1996). This is coupled by the inherent difficulties in making collections from the miners themselves by nature of their informal activity, thus the revenue in the form of taxes or royalties paid on mineral extraction is minimal.

## CONCLUSIONS AND RECOMMENDATIONS

Drawing from the above observations, it can be gathered that the problems that plague the small scale mining industry mostly develop from the lack of appropriate technology. Appropriate technology, unlike any technology has certain basic qualities, adaptiveness, effectiveness and efficiency. While general technological development may be aimed at enhancing effective and efficient production, appropriate technology's adaptive quality ensures its relevance and sensitiveness to the environment within which it is (to be) applied. In this sense, the socio-economic situation of the users is taken into consideration; thus it is supposed to be affordable. Finally, the policy for developing appropriate technology should accommodate an incremental outlook, which allows for constant improvement and adaptiveness to fit in the continuously increasing demand for better and more efficient technology.

However, given the socio-political environment Tanzania's miners operate within, various other considerations need to be incorporated for a feasible technological development program for the small scale sector. These issues are also necessary because they may facilitate the creation of an appropriate environment and responsibility in investing in technology by the miners. The factors to consider include:

— The government should extend the length of claim titles from one to about three years. Legal ownership of claims should be enforced and emphasised;
— Capacity building efforts in terms of training and local forms of organisation should be emphasised. This may include training on appropriate techniques, exposure to appropriate technology, information dissemination, and identification and promotion of local entrepreneurs to produce such technology;

- There is a need to promote access to loans and equipment leasing programmes through the establishment of special credit institutions, while ensuring that they offer interest rate which are affordable and reasonable to artisanal miners;
- Reliable marketing facilities should be promoted;
- A more decentralised mining administrative and monitoring system may be more appropriate in addressing the problems of the miners in their various diverse locations.

## REFERENCES

Chachage, C.S.L. (1995). 'The Meek Shall Inherit the Earth but not the Mining Rights: The Mining Industry and Accumulation in Tanzania in Liberalised Development in Tanzania': in *Studies on Accumulation Processes and Local institutions* by Peter Gibbon (ed) 1995. Nordiska Afrikainstitutet, Uppsala, Sweden.

Cooksey, B. (1994). "Who's Poor in Tanzania? A Review of Recent Poverty Research". REPOA Special Paper 3, Dar es Salaam. DUP.

Gibbon, P. (1995). 'Mechanisation of Production and Privatisation of Development in Post Ujamaa Tanzania: an Introduction in Liberalised Development in Tanzania': in *Studies on Accumulation Processes and Local institutions* by Peter Gibbon (ed) 1995. Nordiska Afrikainstitutet, Uppsala, Sweden.

Hollaway, J. (1993). "Small scale mining in Tanzania: A Technical Assessment", UNDP Project. MWEM, Dar es Salaam.

Jennings, Norman S. (1993). "Small Scale Mining in developing Countries: Addressing Labour and Social Issues". UN Interregional Seminar on Guidelines for the Development of Small-Scale Mining, Harare, Feb. 1993, ILO.

Lwakatare, S. (1993). "Small scale mining in Tanzania: A Study on Institutional Framework", UNDP project No. URT/90/020, Ministry of Water, Energy and Minerals, Dar es Salaam.

Mchaina, D., A.E. Hall and R.S. Svotwa (1987). "Small scale mining in Tanzania: Its significance and the Government's Role", African Mining Conference, Harare, Zimbabwe.

Mutagwaba, W. (1993). "Artisanal Mining Convalescence through Science and Technology", International Conference on Science and Technology for Third World development, Glasgow, Scotland.

Mutagwaba, W., A. Mlaki, and R. Mwaipopo-Ako (1996). Artisanal Mining and Poverty Alleviation in Tanzania, A Draft Research Report. REPOA, Dar es Salaam.

Newman, Chris J. (1987). "Small workings in Zimbabwe: Recklessness or economic Risk?". Conference on Small scale economics and Development, Royal School of Mines, London, England.

Ngonyani, E.A. (1996) "Development of small scale gold and gemstone mining in Tanzania". Paper presented at Prof. J.T. Nanyaro memorial workshop. UDsm 16th January, 1996. Tanzania.

Priester, M. and T. Hentschel (1992). "Small scale gold mining and Processing Techniques in developing Countries". GATE, Vieweg-verlag Braunschweig.

TANDISCOVERY (1996). Baseline Survey of Artisanal and Small scale Mining in Tanzania, a Draft Report, WB/MEM project, 1995.

The Mineral Sector Policy of Tanzania; Draft Paper ESRF - 8/3/1996.

World Bank (1993). "Strategy for African Mining", Mining Unit, Industry and Energy Unit, WB document, Washington DC.

# BIRTH OF SMALL AND TINY BENEFICIATION PLANT

*Shekhar Chakravarty*
Conveyor and Ropeway Services, Calcutta

Beneficiation is complementary to mining. In case of coal, the quality, determined by the amount of ash, plays an important role in its performance and marketability.

For metallurgical coal, the lower the ash content is, the better is the demand. For non-metallurgical coal also, along with other property factors, consistent ash, at different levels, for different industries, are very important, for stable output quality of the Plant and performance thereof too. Also, in case of power industry in India, the input coal quality had seen a gradual decline, over the years, with more and more open cast mining coming up, and per force, boiler designs have been adjusted to poorer coal inputs of around 40% ash, with consequent sacrifice in Plant Load Factor.

A statutory line has recently been drawn by the Ministry of Environment and Forest (MOEF), restricting operation of those Power Plant using (+) 30 per cent ash coal, rendering use of (–) 30% ash coal for Power Plant mandatory from 1st July, 2000. Widespread coal beneficiation, therefore, is going to be a distinct feature in years to come.

In Indian scenario, after a beginning with one or two small privately owned Washing Plants in '50s, a number of big coal washing complexes were built by Hindusthan Steel Limited and National Coal Development Corporation, both in public sector, and subsequently, by the subsidiaries of Coal India Limited.

Raisings in majority of coal mines in India were small and still are, ranging between 300 to 1000 tonnes per day. Simple, economic, flexible, and small pithead application of Coal Washing Units was therefore considered to be more appropriate. The first success in this direction was a 120 TPH Coal Washing Plant at Lodna Colliery of Bharat Coking Coal Limited, Dhanbad, with a new technology from UK, a Modular Barrel-cum-Cyclone Coal Washing Unit using self-generated Slurry as media, implemented by M/s. Conveyor & Ropeway Services under Science & Technology Grant, of Department of Coal. The Wasing Unit is a 3-product one, processing coking coal and producing Clean, Middlings and Rejects. The flow sheet of Lodna Washery is given in Figure 1.

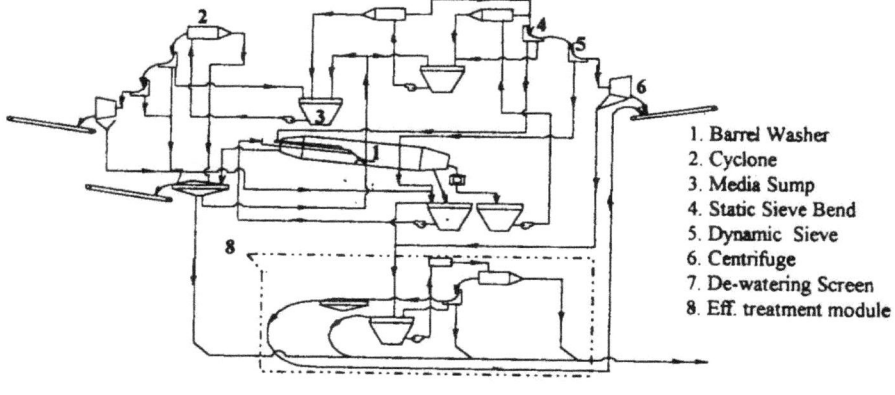

1. Barrel Washer
2. Cyclone
3. Media Sump
4. Static Sieve Bend
5. Dynamic Sieve
6. Centrifuge
7. De-watering Screen
8. Eff. treatment module

Figure 1

The Washing Plant at Lodna is a low profile structure provided with raw coal and products handling equipment, within the limited land available in the colliery. The Washery was designed to be fed by two closely located mines. Attractive feature of this Washing Unit were the following, which prompted the Standing Scientific Research Committee (SSRC) to recommend the S&T grant for its implementation, on commercial basis.

1. Substantially low capital cost, Rs. 50.00 per tonne of feed coal compared to more than Rs. 300.00 prevalent then, on annualised basis.
2. Low operation cost, and limited manual labour.
3. Low profile.
4. Low power and water consumption.
5. Being pithead, infrastructural investments avoided.
6. No foreign media required.

Although the small unit was a success and gave highly promising results, high investment culture of CIL with large Central Coal Washing Plants, unfortunately, did not encourage propagation of the benefits of this small Modular Washery, or further development thereof.

Consumers in the private sector, beset with quality problems took the cue, and a few Modular Washing Units on the same technology of Barrel-cum-Cyclone Unit came up, both for coking and non-coking coal, the smallest one was for 25 TPH. Cost of washing in all these cases could be kept quite low, within Rs. 30.00 per tonne of feed coal.

Quality of coal supplies to small consumers, like cokeries, who need consistent quality of inputs, being indifferent, deliberation with the small users revealed that, even much smaller Washing Units of around 15 TPH could ideally suit the requirement of cokeries to adequately serve the Foundry and other industries for their desired level of ash in hard coke.

A needbased design was developed utilising the technology of the aforesaid Modular Coal Washing Unit.

In Barrel-cum-Cyclone Washer, the Barrel does the work of primary deshaling, whereafter the floats from the Barrrel are further washed in stages of special Cyclones, with self-generated slurry media. The specific gravity of the media is suitably adjusted to derive the desired ash level in product. The Barrel with its drive, support structures etc., however, contributed to major portion of the total cost. Examinations also revealed that pure shales in raw coal constituted only 3 to 4% of raw coal.

There were various discussions with the Cokery owners. Owners Association also organised a full-scale deliberation. The objective was to reduce the size of the Washing Unit, and at the same time make it less investment oriented.

It was these deliberations, which revealed that manual removal of the shales (bands). which is invariably resorted to by all cokeries at their premises, was de facto the deshaling operation, and all owners agreed to live with it, if it meant sizeable reduction in their investments for installation of a small Washing Plant, giving the desired ash level in Clean Coal for production of Foundry/Metallurgical Coke. Majority of Cokery owners during the deliberation expressed that a 15 TPH Coal Washing Plant would be a good size for them to beneficiate their daily raw coal requirements, duly deshaled through manual picking. The 15 TPH tiny Coal Washery Unit thus was born through interaction with the users only.

The Cyclone circuit of Barrel-cum-Cyclone Washer was isolated for the mini washery.

The S&T Plant at Lodna had adequately established that from 30–32% ash raw coking coal, Clean Coal of 17% ash could be produced after deshaling the raw coal by approximately 4 to 5% ash. Assumptions for the Cyclone Plant to be able to reduce ash, from 30–32% feed after deshaling was considered to be quite practical. The aim was to achieve 20–22% cleans for production of Foundry grade coke. And it worked.

A very compact and unitised design for the 15 TPH was developed, which not only could produce 3-products, namely Clean, Middlings and Rejects, but also accommodate itself completely with its Slurry Ponds within the limited spaces that are available within the boundaries of most of the cokeries. Flow Diagram of a typical 15 TPH Plant is Illustrated in Figure 2.

The unit consisted of the following operation :

1. After manual picking of shales crushing of raw coal to (–) 12 mm size.
2. Conveying through Belt Conveyors and storing in a small surge hopper, which is part of the Washing Plant.

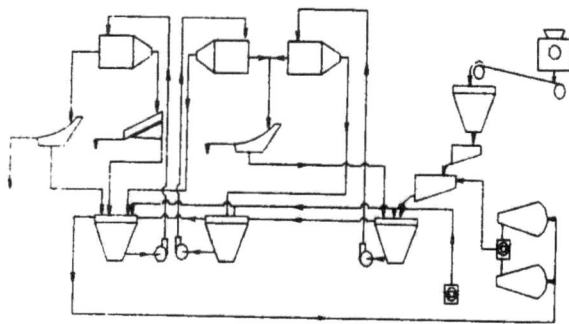

**Figure 2**

3. Take feed from surge hopper into Slurry Tanks and process, first, through 2-stage Cyclone circuit to extract the Cleans at desired ash level, which could be conveniently controlled by (i) adjusting the specific gravity of the Slurry Tanks, and (ii) adjusting the Cyclone spigot under on-running conditions.

   The overflow of the first 2-stage Cyclones formed the Clean Coal, and the underflow of the second stage cyclone constituted the feed for

**15 TPH Coal Washing Plant**

the Middling Circuit. From the Middling cyclone separation, the overflow constitutes the Middling and the underflow the Rejects.

As for quality of products, a few trials varying the static specific gravity of the (–) 0.5 mm Slurry from various Cyclone feed Tanks, and also the spigot openings on Cyclone, the desired product ash, particularly that of Clean and Middling were arrived at.

The layout of a Washing Unit in a typical Coke Manufacturing Complex is illustrated in Figure 3.

The overflow fine slurries from various Slurry Tanks are led to suitable size ponds, from where the overflow water containing very small fine is pumped back to the circuit. This avoids any effluent discharge.

The total investment on account of such tiny Modular Coal Washing Unit of 15 TPH capacity presently comes to approximately Rs. 50.00 lacs inclusive of infrastructural preparations. The capital cost per tonne of raw coal, on the basis of present market price, on this Plant comes to approximately Rs. 60.00/tonne of raw coal, on annualised basis.

Most of hard coke manufacturing units in the coalfields have to live on medium coking coal having relatively high ash around 30–34%. After blending with low ash (approximately 6%) high sulphur Assam Coal and also some good quality Slurry Fines from the Washing Plants in Jharia Coalfield, the average ash in the blend is brought down to 26–27% and Grade-I coke is produced with ash in the end product ranging between 32–34%. This Grade-I coke being used mainly for non-metallurgical use, have marginal value addition. With incorporation of a Coal Washing Unit and following extraction of better lower ash coal, as clean, around 20–22%, foundry grade coke could be produced giving a good value addition. A harmonogramic chart is given in Figure 4.

This would show how this mini Washing Unit could benefit the Hard Coke Manufacturers, and at the same time help better availability of hard coke suitable for Foundry use.

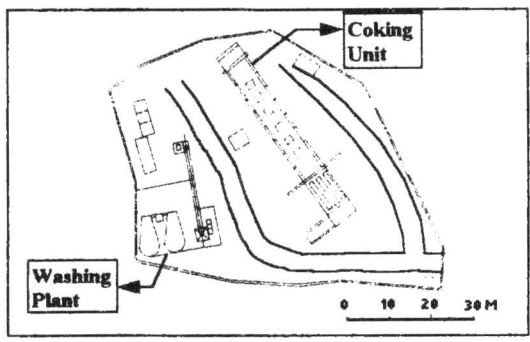

Figure 3

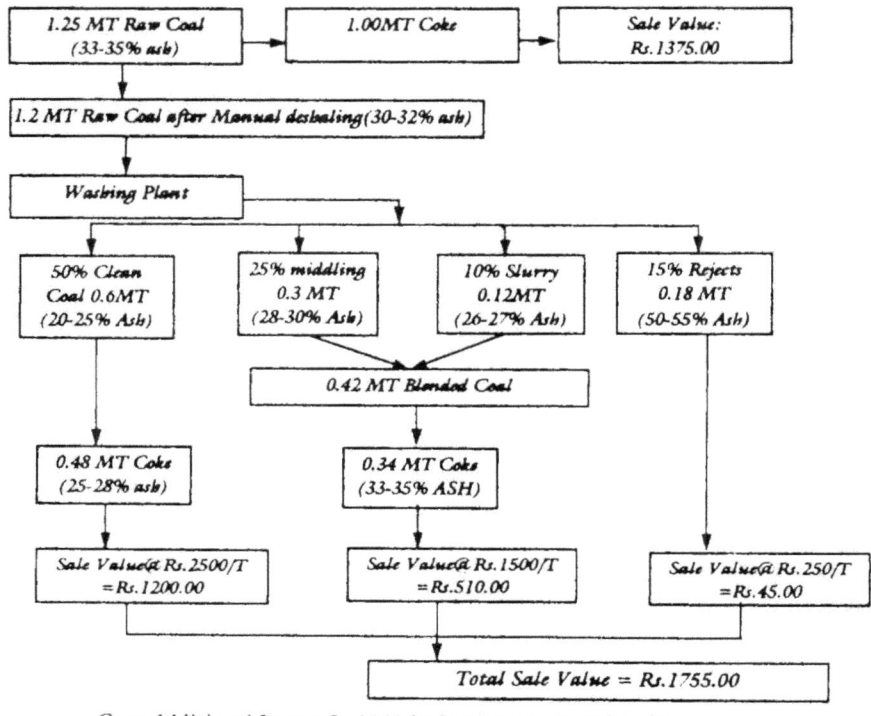

**Figure 4**

The birth of a tiny coal beneficiation unit in India thus has taken place essentially out of compulsion because of non-availability of desired quality of coal by small consumers, and more so, because the small investment on its account did not inflict an undue financial strain on the company, rather resulted in both qualitative and economic gain.

## DESILTING OF LIMESTONE

It would be interesting to note that while the Cyclone section of Barrel-cum-Cyclone Washer technology could serve the users in Hard Coke Manufacturing Unit, the Barrel Washery could be effectively deployed in desilting of limestone smalls from a Limestone Quarry, which are segregated in Crushing process and disposed off as rejects. Because of nearly 50% silt, particularly, during rainy season working, not only their disposal became problematical, but because of the associated silt a good amount of good quality limestone had to be rejected, which on its washing and reduction of the silt to almost 3% could be used as raw material, and thus, optimisation of raw mineral could be achieved. The arrangement and flow sheet of a Barrel Washer for desilting (–)12 mm limestone is given in Figure 5.

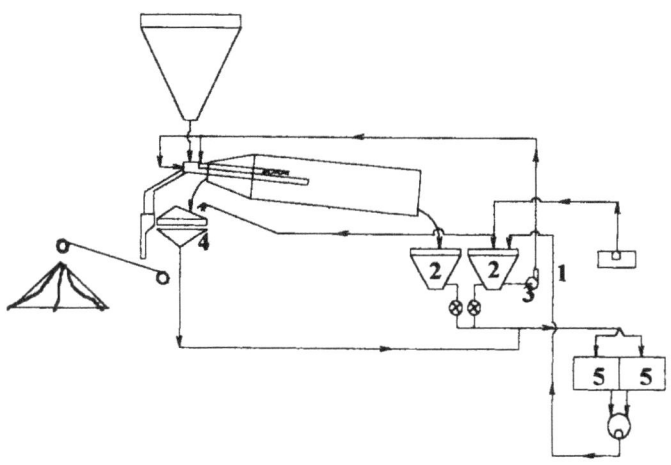

Figure 5

Enquiries have revealed that a number of Cement Plants in their raw material handling are beset with this problem of presence of silt in smaller sizes. Success on the washer at Diamond Cements open up good possibilities for the mines belonging to other Cement Plants.

The first Modular Barrel desilting Unit installed at M/s Diamond Cements, in Damoh, M.P. is of 60 TPH capacity, involving a capital investment of approximtely Rs. 35.0 lacs only.

# SOFTWARE AND HARDWARE REQUIREMENTS FOR SMALL SCALE MINING SECTOR

*A. Santha Ram*
Senior Mining Geologist, Mining Research Cell, Indian Bureau of Mines, Nagpur, India

**INTRODUCTION**

Small scale or Medium scale mining is gaining important position in the mineral sector which has got a potential growth and future. There are many small mineral deposits in our country including base metal deposits, which can be mined, as a cluster of small mines without developing much infrastructure facilities. Small scale mining offers excellent opportunities through in-house training, for transformation of unskilled labour to skilled labour and semi-skilled labour and in this way contribute to the basic skill formation.

Small scale mining, therefore is an important agent or catalyst for economic development. This sector provides a large cash, economy, within short gestation period. The operation of small scale mines has several peculiar attributes which large scale mining sector is not endowed with:

1. The sector has the ability of utilising small and other unexpliotable minerals which are not of much interest for large PSU's (Public sector undertakings) and large private mines.
2. The sector can absorb less skilled/unskilled labour in rural areas and uses scarce and low cost capital efficiency.
3. Small scale mines require modest infrastructure for operation. This sector has also opportunities for indigenous development and advancement and hence contribute to human resource development.

Though the frequent usage of scale concepts for defining the small scale mining in quantitative terms does not exist, generally definitions are based on one or all of the following criteria which exhibits distinct variances as a result of scale of operations: These include labour output, turnover, degree and extent of mechanisation, labour productivity, size of lease and reserves. Most of the small/medium mines were located in remotest places without any proper communication or transportation facilities. Despite these constraints, the small mines generate a substantial employment to the local populace.

## NEED FOR COMPUTERISATION IN SMALL MINING SECTOR

As the information technology in mining sector is fast growing with reference to globalisation, a need has arisen to computerise the small scale mining activities for developing proper database management systems for each mineral/varieties of minerals produced. The basic data collection procedures for small mines need to be carefully planned as this sector is a very important nerve center to the mining industry. It cannot be viewed in isolation, because small mines may transform to medium/large mines. Since most of the small mines do not possess adequate exploration data, it is therefore difficult to assess the future status of these mines in terms of technical viability and productivity.

Thus the small mines demand much attention, because they lack qualified technical staff and the mine owners cannot afford to attract them. Secondly, due to lack of infrastructure facilities at the site (there are many small mines not well connected by roads or having proper communication), these mines could not attract educated/trained personnel to work for longer spells. As a result valuable scientific data could not be collected during operation cycle of the small scale mine. It is the similar case with many small mineral deposits, where proper records on mine geology, lease details, reserves, grade of mineral produced, nature of occurrence, types of host rock and its association etc. were not available, nor even maintained. Data generation, acquisition methods were very poor, because the purpose is not well understood. Now a stage has been reached to computerise the available data for each mine or cluster of mines in a proper format to accomplish this task.

## STATUS OF COMPUTERISATION IN MINING INDUSTRY

Till date, the concept of technical computing is very much confined to large public sector mining companies, which are having their in-house facilities. Unfortunately, most of the companies are either using imported canned software or modified software to meet their requirement. It is, however, felt by the user groups that there are several operational problems associated while using the imported software because of data input limitations. They are not usually suitable to model our type of deposits, especially with reference to method of working, planning and production scheduling. This is due to several unforeseen circumstances and constraints, the planning process cannot be effective to obtain the desired results. Thereby, computerisation is still not effective for obtaining valid data on investment decisions.

## LEVEL OF COMPUTERISATION IN SMALL SCALE MINING SECTOR

The level of computerisation in small scale mining sector is rather insignificant and some efforts are being made. Though PC-based systems are very popular and easily affordable, there is a common fear and doubts regarding its usage in the mining environment. This is due to limited exposure among the mine owners as well as policy makers regarding the utility of computers in small mines. The concept of computerisation at mine site can be achieved by providing basic education/training to the mine owners/operators regarding the possible uses/applications of computers.

To optimise the productivity, some source of financing facility should be made available to small mine or alternatively to a cluster of mines in the nearby areas to have a common computer facility to update the information generated from each mine/mines, which can act as a data collection centre.

The areas for computerisation in small mines include:

1. Manpower records: Summary of personal history statement etc.
2. Payroll Accounting:
3. Inventory Control
4. Maintaining equipment information
5. Data on production from different pits, stock, dispatches, (daily/weekly) gradewise break-up, rejects and its quantity etc.
6. Dimensional information of working or abandoned pits: Monthly measurement of advancing pits (length, width, and depth factors).
7. Planning and production.

The programs useful for small/medium scale mines are:

- Ore reserve estimation
- Grade-tonnage prediction
- Spread sheets and RDBMS
- Gridding and Contouring
- Basic statistics and forecast

Most of the software developed in the academic institutions/higher research organisations will not be applicable for small mines sector and therefore the software problems addressed to the small mines cannot be visualised by the large mechanised mines.

Therefore, small and utility oriented programs are needed. Some of the programs discussed below were developed by the author which were tested under the field conditions and found to be satisfactorily giving meaningful results.

## ORE RESERVE ESTIMATION PROGRAM

For performing the ore reserve calculations of any size and shape of deposit a simple computer program has been developed in FORTRAN machine language. The bench wise ore reserve estimates along with its stripping ratios can be computed. The data input comprise of X and Y coordinates generated for each profile of geological cross section, bench height, bench width, specific gravity values of ore, waste and sectional intervals. At any time any number of geological cross-sections with varying scale factor can be inputted. The acronyms used for transfer of data as input include:

UFX: Upper footwall contact
UHX: Upper Hangwall contact
UEX: Upper Extension
BFX: Bottom Footwall contact
BHX: Bottom Hangwall Contact
BEX: Bottom Extension

The program recognises the acronyms used and accordingly computes the cross-sectional areas of adjoining sections and calculates benchwise the footwall waste, ore and hangwall waste first. Later, it computes the total footwall waste (FWAST), Total ore (TOTORE) and total hangwall waste (TOT HWAST). The bench wise stripping ratios are printed in the output listing for a desired bench height and bench width. The program has an option for computing stripping ratios for different bench heights and widths for intermediate sections as well for intermediate benches. A typical output is shown in Table 1.

## WEIGHTED AVERAGE GRADE PROGRAM: BLOCKWISE

For estimating the benchwise in-situ grade, tonnage and ROM (Run-of mine) grade and tonnage the weighted average grade program provides the solution. This program is developed in FORTRAN machine language, and the data input required for the program include drill hole or blast hole assay values for each hole, bench height, average width of the ore body, specific gravity value of ore and waste. The likely or actual dilution factor from footwall and hangwall can also be inputted in terms of percentages.

Table 1: Output Listing of Summary of Ore Reserve Estimates

| SN | BEN NO | UFX | UHX | UEX | BFX | BHX | BEX | FWAST | ORE | WASTE |
|---|---|---|---|---|---|---|---|---|---|---|
| 8 | 2 | 8.40 | 12.50 | 20.00 | 10.20 | 14.10 | 22.30 | 10.80 | 48.00 | 94.20 |
| 8 | 3 | 10.20 | 14.10 | 22.30 | 11.30 | 15.20 | 24.40 | 6.60 | 46.80 | 104.40 |
| 8 | 4 | 11.30 | 15.20 | 24.40 | 12.20 | 16.10 | 26.40 | 5.40 | 46.80 | 117.00 |
| 8 | 5 | 12.20 | 16.10 | 26.40 | 13.00 | 16.80 | 28.80 | 4.80 | 46.20 | 133.80 |
| 8 | 6 | 13.00 | 16.80 | 28.80 | 13.70 | 17.60 | 31.20 | 4.20 | 46.20 | 153.60 |
| 8 | 7 | 13.70 | 17.50 | 31.20 | 14.40 | 18.50 | 33.50 | 4.20 | 48.00 | 171.60 |
| 8 | 8 | 14.40 | 18.50 | 33.50 | 15.40 | 19.30 | 36.50 | 6.00 | 48.00 | 192.00 |
| 9 | 2 | 8.80 | 12.80 | 19.60 | 10.50 | 14.40 | 21.80 | 10.20 | 47.40 | 85.20 |
| 9 | 3 | 10.50 | 14.40 | 21.80 | 11.30 | 15.30 | 24.00 | 4.80 | 47.40 | 96.60 |
| 9 | 4 | 11.30 | 15.30 | 24.00 | 12.80 | 16.80 | 26.20 | 9.00 | 48.00 | 108.60 |
| 9 | 5 | 12.80 | 16.80 | 26.20 | 14.00 | 18.10 | 28.40 | 7.20 | 48.60 | 118.20 |
| 9 | 6 | 14.00 | 18.10 | 28.40 | 14.90 | 19.00 | 31.10 | 5.40 | 49.20 | 134.40 |
| 9 | 7 | 14.90 | 19.00 | 31.10 | 15.70 | 19.20 | 33.40 | 4.80 | 45.60 | 157.80 |
| 9 | 8 | 15.70 | 19.10 | 33.40 | 16.40 | 22.30 | 34.60 | 4.20 | 56.40 | 165.00 |

The following results pertain to the Bench height of 3 meters

| Bench Number | Total F Waste (MT) | Total Ore (MT) | Total Waste (MT) |
|---|---|---|---|
| 2 | 238.30 | 1003.80 | 2052.00 |
| 3 | 135.00 | 987.00 | 2290.00 |
| 4 | 148.50 | 991.20 | 2569.00 |
| 5 | 126.00 | 987.00 | 2893.50 |
| 6 | 103.50 | 991.20 | 3312.00 |
| 7 | 99.00 | 991.20 | 3757.50 |
| 8 | 121.50 | 1066.80 | 4117.50 |

Grand Total of FWAST : 972.00 MT
Grand Total of Ore : 7018.20 MT
Grand Total of Waste : 20992.50 MT
Waste to Ore Ratio : 1: 2.99

The program first recognises the hangwall and footwall contacts with reference to ore zone, and subsequently the location of drill holes or blast holes as per the bench plan. Then it performs the weighted average grade computations for in-situ grade and tonnage. By applying given percent dilution criteria, it computes weighted average for ROM grade and tonnage. If the user is interested in sub-block grade and tonnage's, the whole bench will be divided into several sub-blocks of specified geometry and accordingly grades and tonnage can be computed. The program can handle 99 nos. of holes with 8 nos. of chemical variables. A typical sample output is shown in Table 2.

**Table 2: Output Listing of Weighted Average Grade**

| | |
|---|---|
| Subblock (SBLOK) | : 1.0 |
| SUBBENCH (SBEN) | : 3 |
| Average Width of Ore Body | : 10m |
| Assay Values of Blast Holes | : 2 (8 Values per hole) |
| Input Assay Values | : 24 |

Results:

| | |
|---|---|
| WTMFAD | : 61.63% |
| TOT. TONAGE | : 2520MT |
| AV. DEV | : 5.67 |
| 10% DILUTION GRADE | : 55% |
| 10% DILUTION TONNAGE | : 2520MT |
| 5% DILUTION GRADE | : 59% |
| 5% DILUTION TONNAGE | : 2645 MT |

## CONTOURING AND GRIDDING

The gridding and contouring program generates contours for geological maps. The data include X, Y and Z coordinates generated in the field. The data can be of irregular spaced or regularly spaced. The gridding algorithms are used to select grid intervals and neighborhood points are joined using inverse square method.

Small scale mine maps can be prepared by using this program with conventional 24-pin dot matrix printer. The contours can be smoothened further to meet user requirements. A typical output of contour map is shown in Figure 1.

## HARDWARE REQUIREMENTS

The hardware requirements for small mines should be inexpensive, versatile, and easy to install and operate. The hardware list includes one or two Personal Computers (PC's), preferably a standard 486-DX2 configuration with dual floppy drives, a colour/monochrome monitor, a standard bi-lingual keyboard (Hindi and English) and a good quality CVT and 24-pin letter quality (LQ) printer. The PC should preferably be housed in a dust-free environment.

## SOFTWARE REQUIREMENTS

Generally, the software requirements depend on the user applications. A proper planning of software requirements need to be identified, with the matching hardware requirements. In other words, the software must be compatible with the existing hardware.

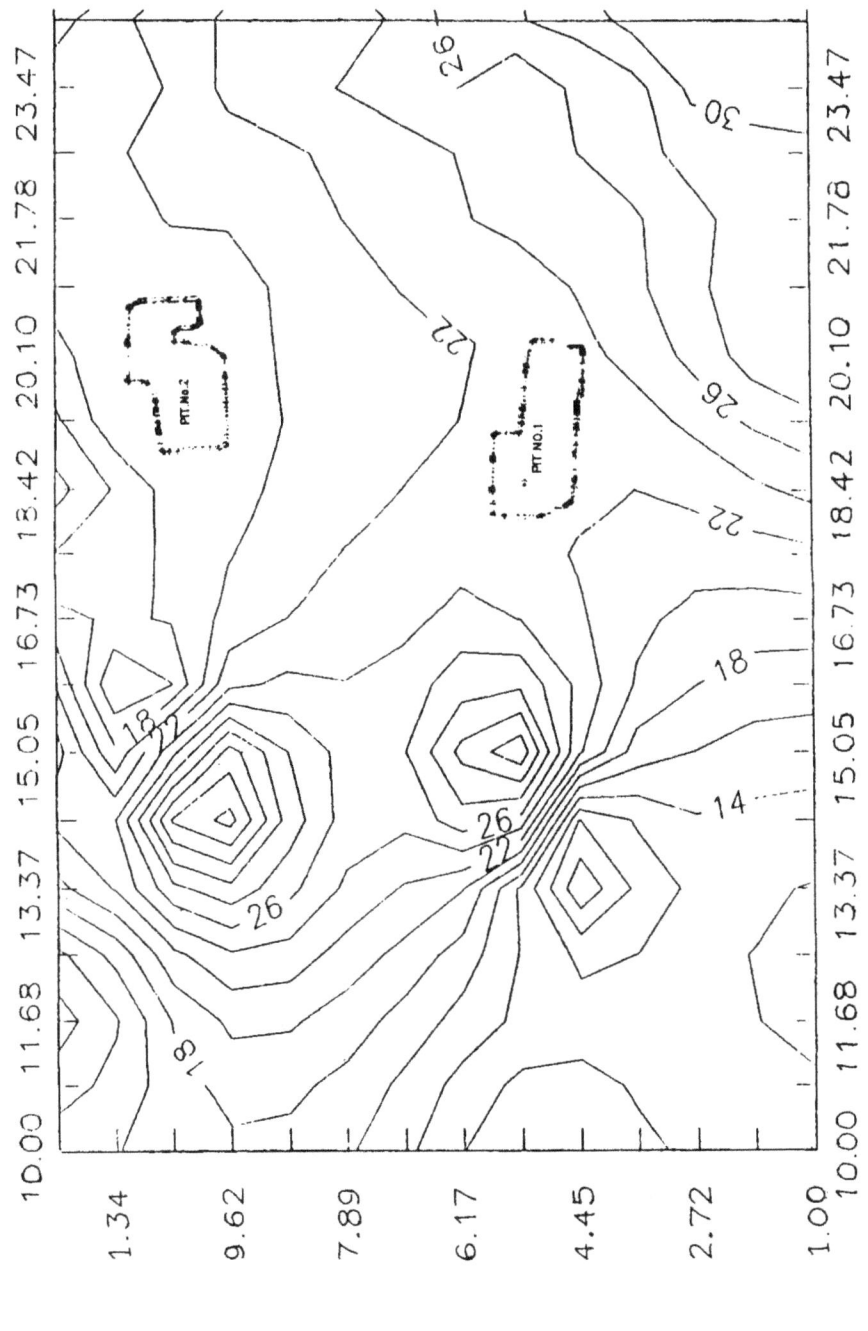

Fig. 1. Contour Map of a Small Mine Area

The software requirements can be broadly categorised into:

a. Utility or General purpose software
b. Application or technical software

The utility or general purpose software include identification of specific software for performing routine office oriented tasks. Usually the software can perform general correspondence, collection, updating, retrieval of data, filing of applications to Government agencies, personal records of employees, payments, budgeting, accounting, quality management, sales, dispatches etc. All these above-mentioned tasks can be accomplished by using "User-friendly" packages, which run on "Window"-based platform. An example of such integrated package has been developed in window-based FOXPRO program for application in small scale mining sector.

The technical software requires flexible and easy to operate systems. Most of the small/medium mines are manually operated and the degree of mechanisation is very low. Therefore, custom made software will be useful to suit the specific requirements. This can be achieved by identifying the nature of operations involved and required data inputs for obtaining the desired results.

Data documentation should be explicit, and neglect in early stages may lead to invalid decisions or sometimes invalid decisions are unchallenged. A number of people become involved during planning and exploitation phase of the mine. They may leave or retire from the typical evolution process. Thus the corporate criteria for investment decisions on small mine may change with time. Therefore, the evaluation must respond to circumstances which change during time taken for carrying mining operations.

## ANTICIPATED GAINS

The anticipated gains while using the utility and technical software are manifold, with increased productivity. Firstly, they are user-friendly and require some amount of basic training skills to operate the machine. Secondly, the user will find more attributes in the software to further refine the data validation, manipulation and reporting.

## CONCLUSIONS

Most of the small/medium scale mines are labour intensive and the degree of mechanisation is relatively too low. Due to inadequate infrastructure and lack of skilled and educated manpower, data acquisition systems for small mines is becoming a difficult. Therefore, small mines require some level of computer culture, which can be introduced through training and developing computer skills.

Thus, the small mines will be in a position to generate valuable databases for production planning and environment, which will help in investment decisions. In majority of small mines, prospecting and exploration data regarding the orebody characteristics is not properly documented to gauge the commercial worth of the deposit. Therefore, proper documentation of data on prospecting/exploration, grade-tonnage characteristics will aid in better understanding of the deposit potential. The development of computer culture in small scale mining sector is a challenging task, which requires motivation, hands-on experience and constant encouragement from the management to fulfill the desired objectives.

## ACKNOWLEDGEMENTS

The author wishes to thank Shri O.P. Sachedeva, Controller General, Shri M. Mukherjee, Chief Controller of Mines, Shri A.N. Bose, Controller of Mines (Central), Indian Bureau of Mines, for guidance and granting permission to present the paper. The author also wishes to acknowledge the valuable guidance received from Shri K.L. Jangida, Regional Controller of Mines, Mining Research Cell, Indian Bureau of Mines during the preparation of the paper.

### REFERENCES

1. K.D. Sharma (1976). Programming in FORTRAN IV, Affiliated East-West Press, New Delhi
2. Microsoft FOXPRO "Relational Database management system for Windows", Microsoft Corporation, USA.

# SECTION 4

# Sharing Experiences World Wide

# PROSPECTS AND PROBLEMS: SMALL-SCALE GOLD MINING IN PAPUA NEW GUINEA AND THE PHILIPPINES

*Don Stewart*
International Development Technologies Centre, University of Melbourne, Australia

**INTRODUCTION**

In both of the countries discussed in this paper the mining industry is a significant export earner and gold mining is a major part of that industry. Both are in the list of the top ten world gold producers. In the case of Papua New Guinea mining is also a major source of government income. Because the amounts of capital invested in mining are often very large by developing country standards, there has frequently been an expectation that mining would also stimulate more generalised economic development and thus achieve broader development goals. In this regard there has often been some disappointment; large mines tend to operate as 'enclave industries' within developing countries (Emerson, 1982; O'Faircheallaigh, 1984; Cobbe, 1979) and the multiplier effects have not been as great as might have been expected from the size of the enterprise.

The mining developments which are usually thought of when we speak of the mining industry in developing countries have characteristically been large-scale, capital intensive and foreign owned. These characteristics tend to be interrelated and the question we will address in this paper, looking at Papua New Guinea and the Philippines, is whether parallel promotion of the converse, small-scale, labour intensive, indigenous mining might aid in meeting the more generalised development objectives not achieved by large-scale mining. There have been a number of papers written examining the importance (Carman, 1985) and scope (Wels, 1983) of existing small-scale operations in developing countries and also assessing in qualitative terms their advantages and disadvantages (Gocht, 1980; Carman, 1979; Stewart, 1989). There have also been several conferences devoted to small-scale mining (Meyer and Carman, 1978; Neilson, 1980; Blowers, 1983; anon. 1984).

A paper by Noetstaller (1984) has examined the above question from the viewpoint of economic theory and the general conclusion is that a mining development which does not match a country's comparative advantages in factor endowments (this for developing countries often means among other things an abundance of cheap labour), will not contribute in an optimum way to the economy of that country. Even if capital intensive and labour intensive mining projects both show similar profitability in a financial sense, the Net National Value Added will typically be greater for the labour intensive projects and also, because of the smaller front end investment, they will start to generate funds for national development at an earlier date. Also more of the Net National Value Added will be in the form of income for workers, thus contributing more directly to the raising of living standards.

Of course, nature has the major say in the technology to be employed in the case of a particular ore body. Some orebodies such as large low grade porphyry copper deposits are only amenable to large-scale capital intensive methods but, it is equally true that some patchy deposits are ideal for small-scale labour-intensive mining. Information comparing the use of labour intensive and capital intensive construction methods in developing countries (Edmonds, 1981; Scott et al., 1983; Scott, 1986; Scott Wilson Kirkpatrick and Partners, 1978) show that, although there are certain operations in the construction industry which can be economically carried out by labour-intensive methods in developing countries, transport of material over any distance is best done using machinery and the performance of labour in excavation is much more variable than that of machinery because of the effects of ground conditions, weather etc. Thus on the grounds of economic efficiency, as well as humanitarian grounds, labour-intensive mining should be limited to small-scale operations. Also comminution, the large energy consumer in mineral processing, is not really suitable for manual methods. This is why alluvial mining is a suitable candidate for labour intensive methods. Nature has already largely carried out the comminution process. The relationship between size of deposit, grade and feasible scales of mining and capital/labour intensities is discussed in more detail elsewhere (Stewart, 1989).

As well as these technical and economic issues, there are other aspects such as social and environmental considerations, which we should also address, if we are interested in development in its widest sense.

## GOLD MINING IN PAPUA NEW GUINEA

Gold mining is not an indigenous activity in Papua New Guinea. It first started in the country in 1888 at Misima Island and developed further with gold rushes in Morobe Province in 1926 which developed into the manganoquartz open pit mining in the Wau region and the dredging operations of the Bulolo Valley, which recovered 66 tonnes of gold in the years 1932 to 1964.

The commencement of production at Bougainville Copper in 1972 was the start of a new era in mining in Papua New Guinea. The mine at start-up was seen as a copper mine, with by-product gold but with the subsequent rise in gold prices gold has become an equally important co-product and Bougainville Copper was the country's largest gold producer from its start-up until overtaken by the Ok Tedi mine and its subsequent closure due to rebel action. Ok Tedi commenced production in 1982 and is now one of the world's largest gold producers. As well as these very large-scale gold producers, there was for many years a smaller mine at Wau (New Guinea Goldfields) which had been in operation since the early thirties. Another small mine, Clarkes Deposit at Mount Victor near Kainantu commenced operation in 1987 but both this mine and the one at Wau closed during 1990. New mines at Porgera and on Lihir and Misima Islands have also commenced operation recently and some ninety gold prospects are currently at various stages of evaluation in the country.

The above represents the modern mechanised sector of gold production in Papua New Guinea but in the Mines Department Report of 1949 we read: "During the year a party of natives commenced mining on their own accord on the Wanion River . . . . This is the first time in the history of the goldfield that natives have produced gold on their own behalf." Indigenous small-scale mining has continued to grow since that time and it is estimated that at present there are over 5000 people involved (excluding the gold rush at Mount Kare). The techniques used in these operations have been described elsewhere (Blowers, 1985; Stewart, 1987). In a survey of miners in the Wau/Bulolo region (Blowers, 1985) it was found that generally each lease is worked by a group of miners, normally kinsmen of the leaseholder. On average, each site supported 6 to 7 workers using 2 to 3 sluice boxes and about every second site used a small portable water pump. Mining lease areas are often on the miners own tribal land but other land may be worked on a tribute basis at a cost of 5–15% of gold earnings. Simple pick and shovel methods are used to pass 2–3 cubic metres a day of gravel through the sluice boxes. Miners are usually working ground containing over 1 to 2 gram of gold per cubic metre and mostly recovering somewhere between 50 and 300 gram per month. By using information on gold sales by National miners (Canavan, 1983) and information on the cost structure of a typical small-scale mining operation (Stewart, 1983) it is possible to compare the contributions of this activity to incomes of citizens and the government as a proportion of contribution to GDP. These values can be compared with those calculated by Emerson (1982) for the large mines, Bougainville and Ok Tedi. Such a comparison is shown in Table 1.

Table 1: Comparative Benefit to Papua New Guinea from Foreign-owned Mining Operations and Small-scale Miners (1982)

|  | Disposable income of nationals (% of contribution to GDP) | Public Sector income (% of contribution to GDP) |
|---|---|---|
| Bougainville | 14.3 | 36.1 |
| Ok Tedi | 21.4* | 3.0 |
| Small-scale national miners | 70.5 | 13.9 |

* This figure was obtained during the construction phase of the mine and is higher than that which would apply under normal operating conditions.
Source: Stewart, 1987; Emerson, 1982.

Although most miners have been using the same simple manual methods for many years, in the last two years there have been some indications of evolutionary development of indigenous mining. Equipment salesmen have belatedly started to take an interest in the small-scale local miners and some miners have purchased manufactured metal sluice boxes and small suction dredges. In at least one case, a further development on small-scale leases has been the use of hydraulic excavator, trommel screen, sluice and pump for larger-scale mechanised operation.

Another new development on the indigenous gold mining scene in Papua New Guinea during the past three years has been a substantial gold rush at Mount Kare on the border of the Southern Highlands and Enga provinces. This is an area where an overseas mining company (CRA) holds a prospecting licence but where in the period February to May 1988 some 3000 people recovered gold valued at more than $100 million (anon., 1988). By the end of 1988 the numbers involved had risen to 9–10,000 miners. Separation techniques used have been extremely crude as only manual methods, (shovels and panning dishes) are allowed by PNG law on an existing lease, but the gold has been easily won because much of it consisted of sizeable nuggets. In another year the rush was nearly over with the number of miners down to around 300 in October 1989 and at that time an agreement was announced between CRA and the landowners to mine the rest of the gold in the deposit, estimated to be worth $300 million. Under the terms of the agreement the landowners would receive a half share of the gold (anon., 1989).

## GOLD MINING IN THE PHILIPPINES

Unlike Papua New Guinea, gold mining in the Philippines has a history which goes back to pre-historic times. When the Spaniards arrived in 1570 they found extensive placer mining in a number of districts (Paracale, Masbate, Baguio, Surigao and Cagayan de Oro). The arrival of the Americans in the 1890s also meant the arrival of larger-scale modern methods. Dredges were introduced to mine placer deposits and a number of quartz vein mines were also opened up. Apart from the cessation of operations during World War 2 gold mining in the Philippines has continued to grow and in 1985 reported production amounted to some 33 tonnes. In recent years up to half the production has come from porphyry copper mines.

As well as the larger-scale mines, the Philippines has many small-scale gold miners. In 1984 the government recognised the place of small-scale mining in the economy in Presidential Decree No.1899 which laid down guidelines for the regulation of small-scale mining, with the objective of increasing productivity and encouraging wider distribution of wealth from mineral resources. The decree defines small-scale mining as an operation that produces less than 50,000 tonnes per year, requires an investment of less than Pesos 500,000 (US$ 25,000) and has a labour cost to equipment utilisation cost ratio of more than one. The mines which must not be more than 20 hectares are licensed for a renewable two-year period. The usual mining regulations are not applied to small-scale miners but each miner is supposed to submit quarterly and annual reports of operations. Subsequent to the Presidential decree, a letter from the Minister reduced the area for the mine to 2 hectares but the miner was still required to produce a map showing the location of his lease which was difficult for illiterate miners. Later, a Pesos 5 licence for small-scale miners was introduced which gave them the right to mine anywhere in a particular province, providing they have the permission of the surface owner and pay adequate compensation (Balce, 1986).

Small-scale gold miners in the Philippines fall into two main categories. On the one hand there are the alluvial miners, operating in small groups, often as a family, using hand tools and a sluice box for gold recovery in a manner similar to that described for the miners of Papua New Guinea. The other type of small-scale operation is that of the miners working quartz veins. These are generally underground mines. The miners will drive a small adit into a hillside or sink a shaft up to 20 metres deep and dig a drive to intersect the vein and then work upwards to extract the quartz, using mainly hand tools and very few explosives. The ore is batch ground in rod mills, this being the only mechanised part of the operation. It is then manually loaded onto small sluices or strakes and the concentrate from these is treated in an amalgamation mill. In many cases the tailings from the sluices, which may still contain 10–20 g/t gold, are sold to a larger mining operation who treat it in a modern carbon-in-pulp recovery plant.

Organizational structures vary but often a group of miners will operate a mine as a cooperative, with each of say 20 miners having a share and one share each also being held by the financier, who puts up the risk capital, and by the owner of the dewatering pump (the main capital input for the mine). The processing plant may be owned by someone else, sometimes the "landowner" who will charge a fixed fee per can of ore for grinding. In this case use of other equipment and mercury is "free" and the "landowner" will probably also be the goldbuyer.

The relationship between small-scale and large-scale miners can be surprisingly harmonious with some interesting examples of synergism. On one large-sale mining lease near Baguio small-scale miners are operating within the lease boundaries. The mine owners say that the small-scale miners are doing a useful job of proving a portion of the ore body. This is because the large mine buys all their tails so they know the grade being mined and the quantity of gold extracted. In another situation small-scale miners are actually working the mine waste dumps.

There is also evidence of evolution of the technology in the processing of ore from small-scale mining. In the Baguio region some operators have started crude cyanide heap leaching operations. Tailing containing 10–20 g/t are bought from the small-scale miners as the feed for the heap leach plant. The tails from this crude operation still contain around 5 g/t and so can be sold to the large mining company as this is still higher than their cut-off grade for their own mining operations.

Small-scale miners, or gold panners, are found in the Baguio, Paracale and Davao districts in large numbers but also in a number of other parts of the Philippines. Official figures put the number of such miners at 200,000 but private gold traders say there are many more such people operating (anon., 1987a). The most active area at present is Davao del Norte where the Bureau of Mines and Geosciences put the production level at 600 kg a month (Balce, 1986). Gold traders suggest a figure six times this, with trade worth US$ 100 million a month (anon., 1987a). This area has all the hallmarks of a gold rush. The official figures for gold production in the Philippines in 1985 and 1986 are shown in Table 2. Even they show the large and growing importance of small-scale gold mining to the national economy but considerably understate its real production levels because of the large quantity of gold which is traded unofficially.

A number of aspects of small-scale gold mining in the Philippines are causing concern at present. One is the matter of safety. As stated previously, the normal mining regulations are waived for small-scale miners but even if they were not the authorities would be unable to enforce them because of the very large number of operations, particularly in the Davao

**Table 2: Gold Production in the Philippines**

| Producer | Quantity(kg) | | Value (Pesos) | |
| --- | --- | --- | --- | --- |
| | 1985 | 1986 | 1985 | 1986 |
| Primary producers | 9,529 | 9,835 | 1,696,305,799 | 2,211,404,374 |
| Secondary producers | 15,449 | 14,155 | 2,835,700,268 | 3,351,599,420 |
| Small-scale miners | 8,085 | 11,440 | 1,555,872,874 | 2,832,568,940 |

Source: Philippines Bureau of Mines and Geosciences, 1987.

del Norte area. Tunnels owned by different mining groups criss-cross each other with no distinction between leases and cave-ins are common. In one report of a cave-in in 1987 it was thought that 120 people were killed but no proper lists of workers were available, so not even the names and numbers of deaths were known (anon., 1987b). Another aspect of the safety problem is the matter of mercury poisoning. Cases of death from acute mercury poisoning have been reported due to the way in which miners distil excess mercury off gold amalgam by heating it in the open (anon., 1988b). There is also the problem of mercury entering the river systems, and subsequently the food chain, via various forms of seafood (anon., 1987c,d,e,f,g). The Bureau of Mines and Geosciences is endeavouring to educate miners in the dangers of mercury and to encourage the use of retorts to distil mercury but this is difficult because the health effects are not immediately obvious. Another environmental issue is the deforestation caused in areas around Baguio and in Davao del Norte by the cutting of trees for mine timbers. Another problem area is that of law and order. In the more established small-scale mining areas the miners themselves have developed effective procedures for dealing with disputes and controlling the area but in Davao del Norte, with government control largely ineffective and a great deal of illegal gold trading going on, it is inevitable that control will pass into the hands of those operating outside the law (anon., 1986a).

## GOLD MINING AS A DEVELOPMENT ACTIVITY

The point has already been made that large-scale, foreign owned, high technology mining operations often make a sizeable contribution to government income in developing countries but as we will see later, we cannot expect very much in this area from small-scale mining and so we must look at its contribution in other terms. One thing which small-scale mining does quite effectively is to get funds into the hands of local people in rural areas. Table 1 illustrates how in Papua New Guinea a much greater proportion of the funds generated by small-scale mining remain in the country compared to the large mines. It is notoriously difficult to find out information about the income of small-scale gold miners but from figures obtained for one small-scale mine in the Bicol Region of the Philippines we would estimate that the individual miners would be receiving an income of approximately six times the minimum rural wage. We suspect that this is typical. Very few miners are making a fortune but the activity is more profitable than alternative rural activities. Small-scale mining also does more than large-scale mining in encouraging other supporting activities in the rural areas. For example, the small-scale miners of Davao del Norte operate some 2000 ball mills which are locally manufactured and power and water for these operations are also supplied by local entrepreneurs (anon.,1986b). We can conclude that small-scale mining is useful in providing income and in generating economic activity in rural areas, thus helping to halt the drift to the cities, something which is a goal of most developing countries.

Another important aspect of small-scale mining is that it can form the basis of a truly local industrialization and development process. We mentioned in the introduction that an industry which does not fit a country's technical, economic and organizational environment will not contribute in an optimum way to the nation's development. This is where small-scale mining succeeds and large-scale mining fails but more than that, small-scale mining can grow and evolve in line with the country's technical, economic and organizational resources and indeed can stimulate that growth to produce a fully indigenous fully integrated national mining industry with strong links to the rest of the economy. We have quoted evidence of this sort of evolutionary growth in both PNG and the Philippines.

We still need to look critically, at the type of rural development fostered by small-scale mining to see how desirable this is. The small-scale gold mining studied in Papua New Guinea (Stewart, 1987) was alluvial mining conducted mainly in river beds and thus did not have an impact on agricultural land and so, unlike cash cropping, did not have the effect of reducing subsistence food production. Also it could be carried out in conjunction with agriculture by the groups involved, so that slack periods in the agricultural cycle could be used for mining and mining could also be readily stopped or started depending on the market prices. While this may be true in PNG, small-scale mining in the Philippines shows severe problems of health and safety and environmental pollution. The situation with regard to mercury use could be considered as critical.

Another aspect of small-scale mining which has often been criticised is that the mining practices of small-scale miners will lead to an inefficient use of the nation's mineral resources. The small-scale miner's interests are short term. He may "pick the eyes" out of a deposit and thus not recover as much value as would be possible. One might also expect that his crude recovery methods would mean that the tailings contain high levels of valuable material. Interestingly, however, the small-scale miners of Papua New Guinea were observed to use a method of operating their sluice boxes with low flow rates of water which laboratory studies (Stewart, 1986) showed give high gold recoveries of fine gold, although at the expense of a low throughput. This is a logical approach to non-mechanised mining. Throughput involves manual labour and so the miners were maximising return on effort invested. The small-scale miners in the Philippines, operating on vein deposits, also use a method of operating their sluice boxes which is appropriate for recovery of fine gold (Stewart and Gayap, 1988) and although their overall recovery may only be 30–40%, because they sell their tailings to large mines operating cyanide leach plants, the overall recovery of the nation's mineral resources is probably quite reasonable.

The above evaluation applies particularly to what might be described as "steady state" small-scale gold mining but there are certain additional features present in gold rushes like those occurring at Mount Kare in PNG and Mount Diwata in the Philippines where the value of the gold recovered would seem to be of the order of hundreds of millions of dollars in each case. In both cases there is evidence of conspicuous and wasteful consumption in the way of gold rush towns of the past and also considerable violence and lawlessness. It is also very likely that they draw people away from agriculture and other productive rural activities. While it would be easy to see this as wholly undesirable in the context of national development, it is also necessary that we remember that goldrushes of the past in places like the US and Australia were the forerunners of a prosperous and comparatively egalitarian society (Blainey, 1969). It could be argued that although wasteful consumption is very much in evidence many more participants may be quietly saving their proceeds to be invested later in constructive activities. Again, although it may at first seem paradoxical, gold has been referred to as "the democratic mineral". In the early stages of the development of a goldfield all compete on fairly equal terms. Hard work and luck are needed but resources and capital play less of a part than in most other forms of economic activity. Gold diggers tend to be self-reliant and fiercely independent people and this very aspect may play a part in breaking patterns of dependency. Only history will tell whether this aspect proves to be a valuable contribution to the development of Papua New Guinea and the Philippines.

## POLICY ISSUES

Given the lure of gold, particularly when we have deposits as rich as those which exist at Mount Kare and Mount Diwata, it is hardly likely that even the most authoritarian regime could legislate to limit small-scale gold mining. Nor is it likely that governments can hope to use such activities as a source of revenue. Because gold's value is so high in relation to its volume and it is so negotiable, experience shows that smuggling and illegal trading will always circumvent such efforts of government (Walrond, 1986; Holloway, 1984). Hence, even when the total value of the product produced from small-scale mining is quite large, it is unlikely that any government, particularly one in a developing country, can hope to gain substantial revenue from taxing this activity. Governments should rather aim to be in a position where they have reasonably accurate information on the activities of small-scale miners so that they can take appropriate steps to maximise the benefit to the country from such activities and do what they can to reduce the undesirable effects. The following are areas where the government could most appropriately be involved.

1. *Designation*: It is clearly in the national interest that certain areas be reserved for large-scale mining activities but there are other areas where small-scale mining might be the most desirable way of exploiting an ore body and these areas should be designated small-scale mining areas. In general terms, rich patchy or thin vein deposits are particularly suitable for small-scale exploitation.

2. *Exploration*: Small-scale miners generally do not have the skills and resources either to systematically evaluate an area or to make use of the data on mineral occurrences available in a country's bureau of mineral resources and so dissemination of information on worthwhile areas would be useful.

3. *Technical Development and Information*: The methods used by small-scale miners have evolved over many years but have often not been the subject of scientific study. For example, the sluice box has a history which goes back at least two thousand years but the physical processes involved in its operation are not well understood. At the University of Melbourne we have been working on developing such an understanding so that the design of efficient sluice boxes can proceed while maintaining the inherent simplicity of the device.The dissemination of information on appropriate sluice box design and other more efficient recovery methods and safe working practices will benefit everyone involved. The recent publication of a handbook in simple language for the small-scale gold miners of Papua New Guinea (Blowers, 1989) is a good example of what can be done.

4. *Regulation*: The difficulties of regulation under gold rush conditions have been mentioned already but one of the major objectives should be to encourage those forms of ownership which will minimise exploitation of workers and thus encourage equity and safe working practices.

History suggests that self-regulation is likely to prove the most effective approach.

5. *Finance*: Mining is seen by most financiers as a risky business and they tend to be reluctant to invest money in mining, particularly small-scale mining. Some funds are needed, however, even if these are fairly modest, so that small-scale miners who have both a suitable orebody and the technical and entrepreneurial skills can advance their operations to the next level of sophistication and the evolutionary growth of an indigenous mining industry, as discussed earlier, can proceed. It may be that the government will need to intervene in the financial market to introduce a financial institution with an understanding of the small-scale mining industry.

## CONCLUSION

Large-scale capital intensive mining in developing countries provides funds for government activities but is not so effective in achieving more general economic development. Small-scale labour intensive mining is more effective in this regard and, although it has some undesirable features, it is worthy of encouragement, as a parallel activity. There is scope in the areas of technical development, regulation and information dissemination to reduce the undesirable effects while preserving the benefits. The "bottom up" growth of an indigenous mining industry by the evolution of small-mining is likely to prove a much more valuable activity in general development terms than the activities of foreign-owned large mining enterprises however efficient they may be in technical and economic terms.

### REFERENCES

Anon., 1985. Report on the workshop on mineral policy for small-scale mining held at New Delhi, India, Nov.1984. *World Mining Equipment*, March, pp 36–37.
Anon., 1986a. *Sunday Inquirer*, Manila July 20th.
Anon., 1986b. *Manila Bulletin*, Aug. 31st.
Anon., 1987a. *South*, May pp 12–13.
Anon., 1987b. *Manila Bulletin*, Aug. 2nd.
Anon., 1987c. ibid, July 10th.
Anon., 1987d. ibid, July 11th.
Anon., 1987e. ibid, July 23rd.
Anon., 1987f. ibid, July 27th.
Anon., 1987g. ibid, July 24th.
Anon., 1988a. *Melbourne Age*, 14th May.
Anon., 1988b. *Asia Week*, Sept. 9th.
Anon., 1989. *Melbourne Age*, October 3rd.
Balce, G., 1986. Private discussions at Philippines Bureau of Mines and Geosciences, Sept.
Blainey, G, 1969. *The Rush that Never Ended: A History of Australian Mining*, (Melbourne University Press, Melbourne).
Blowers, M., 1983. *Proceedings of the Conference on Small-Scale Mining in Papua New Guinea*, held at Lae, Papua New Guinea. (The PNG University of Technology, Lae, PNG).
Blowers, M.J., 1985. Small-scale mining in Papua New Guinea, *Proceedings of Asia Mining*, pp. 153–171, (Institution of Mining and Metallurgy, London).
Blowers, M., 1989. *Handbook of Small Scale Gold Mining for Papua New Guinea*, (Pacific Resource Publications: Christchurch, New Zealand).
Canavan, P., 1983. Current policy issues affecting Papua New Guinea alluvial mining, *Proceedings of the Conference on Small-Scale Mining in Papua New Guinea*, Lae, PNG, pp. 74–81, (PNG University of Technology, Lae, PNG).
Carman, J.S., 1979. Small-scale mining in the developing world: problems and opportunities; in *Obstacles to Mineral Development — A Pragmatic View*, (Ed. B. Varon), (Pergamon Press: New York).
Carman, J.S., 1985. The contribution of small-scale mining to world mineral production, *Natural Resources Forum*, 9: 119–124.
Cobbe, J.H., 1979. *Governments and mining companies in developing countries*, (Westfield: Boulder, Colorado).
Emerson, C., 1982. Mining Enclaves and Taxation, *World Development*, 10: 561–571.

Gocht, W., 1980. The importance of small-scale mining in developing countries. *Natural Resources and Development*, 12: 7–18.

Holloway, J. (1984). The administration of small gold producers. *The Courier*, 87: 87–89.

Meyer, R.F. and CARMAN, J.S. (eds). (1978). *The future of Small-scale mining, Proceedings of the first international conference*, Jurica, Oro., Mexico, Nov.-Dec., (McGraw Hill: New York).

Neilson, J.M.(ed). 1980. Strategies for small-scale mining and mineral industries, *Proceedings of an AGID Regional Workshop held at Mombasa, Kenya, AGID Report No.8*.

Noetstaller, R., 1984. On the appropriate technology for mining sector projects in different economic environments, *Berg und Huttenmannische Monatshefte*, 129: 385–392.

O'Faircheallaigh, C., 1984. *Mining and Development*, (Croom Held: Sydney).

Philippines Bureau of Mines and Geosciences, 1987. *Mineral News Service*, June.

Scott, D. and Krishman, V., 1983. Appropriate Technology for Fiji, *Construction Papers, Vol. 2, No. 1*, (Chartered Institute of Building: Fiji).

Scott, D., 1986. Choosing the Appropriate Method of Construction. *International Journal of Development Technology*, 4: 205–213.

Scott Wilson Kirkpatrick and Partners, 1978. *Guide to Competitive Bidding on Construction Projects in Labour Abundant Economies*, (World Bank: Washington, D.C.).

Stewart, D.F., 1983. Small-scale mining: an appropriate industry for Papua New Guinea?, *Proceedings of the Conference on Small-Scale Mining in Papua New Guinea*, Lae, PNG, pp11–20,(PNG University of Technology, Lae, PNG).

Stewart, D.F., 1986. Operation of the sluice box under conditions of low water flow, *Bull Proc. Australas. Inst. Min. Metall.*, 291: 81–85.

Stewart, D.F., 1987. Small-scale mining and development: The case of Papua New Guinea. *Natural Resources Forum*,11: 219–227.

Stewart, D.F and Gayap, J., 1988. Unpublished observations.

Stewart, D.F., 1989. Large-scale v small-scale mining: Meeting the needs of developing countries, *Natural Resources Forum*, 13: 44–52.

Walrond, G.W., 1986. Small gold mines — the production and declaration problem: another dilemma for a small developing country, *Natural Resources Forum*, 10: 343–349.

Wels, T.A., 1983. Small-scale mining — Stewart the forgotten partner, *Trans. Inst. Min. Metall. (Sec. A: Min. Industry)*, 92: A19–A27.

# SMALL SCALE MINING AND THE ENVIRONMENT IN ZIMBABWE: THE CASE OF ALLUVIAL GOLD PANNING AND CHROMITE MINING

*Oliver Maponga*
University of Zimbabwe, P.O. Box MP167, Mt. Pleasant, Harare

## INTRODUCTION

Mining is an important industry to the Zimbabwean economy contributing over 7.2 per cent (1994) to Gross Domestic Product (GDP), 5 per cent to employment and over 32 per cent to total export earnings. In addition, mining contributes significantly to fiscal revenue through company and personal income taxes and also acts as a supplier of raw materials to the rest of the economy especially agriculture and manufacturing. Over fifty minerals are currently mined in Zimbabwe with gold being the most important, contributing over 34 per cent to total mineral exports and employing over 31 per cent of the labour currently employed in the formal mining sector. Other important minerals produced in Zimbabwe are; copper, nickel, asbestos, diamonds, chromite, and iron ore. There are over 1,400 mines currently registered as active in Zimbabwe of which gold mines constitute the largest proportion followed by chromite, copper, nickel, diamonds and other base metals.

The mining sector in Zimbabwe can be divided into two broad categories; large scale and small scale operations. Large scale mines constitute the largest group in terms of capitalisation, tonnage, mineral production, turnover and level employment but are a smaller group in terms of number of mines. The small scale mining sector, defined by either output, size of mining concession, annual profits, amount of revenue from sales, capitalisation or employment terms has risen in importance in developing countries in the last few years and Zimbabwe is no exception—see Noetstaller (1993) and Ghose (1993) for various definitions of small scale mines. The United Nations estimates that the small scale mining sector contributes about 10 per cent to total world mineral production. In a way, small scale operations occupy a niche in the mining industry by 'efficiently' recovering minerals from deposits which are too small to support large-scale operations. They are important in employment generation, human resources development, expansion of the national minerals inventory, industrial linkages and in generation of export earnings and revenues through income taxation.

## Overview of Small Scale in Zimbabwe

Zimbabwe is a country with a long tradition of small scale mining (Munyoro, 1993). Small scale mining in this country can be divided into two broad categories; the formal (legal) and the informal mines. Formal mines are those mines which are registered by the Ministry of Mines and operate within the confines of the Mines and Minerals Act Chapter 165 and its other amendments. These mines are normally owned by individuals, small syndicates and/or cooperatives. According to Maponga (1993) four broad categories of formal small scale mines/miners can be distinguished, that is:

- those mines operated by experienced individuals,
- those mines operated by groups of miners using unsophisticated equipment,
- registered gold panners, and
- those mines operated by cooperatives

Chromite, tin and tantalite mining cooperatives fall within the realm of formal mines. Government created an environment conducive to co-operative mining in 1985 after the major chromite ore claim holders — ZIMASCO (Union Carbide) and ZIMALLOYS (Anglo American Corporation) had decided to restructure their chromite operations and reduce costs following a glut in the base metals market. As co-operative mining fell within the government's development agenda, structures were created to facilitate the development and growth of this sub-sector. The Zimbabwe Mining Development Corporation (ZMDC) was tasked with helping co-operators, supervising their operations and ensuring that safety and mining regulations were adhered to at all times. A trial project at Ngezi Mine on the Great Dyke in 1985 heralded the birth of co-operative mining in Zimbabwe. The sector has grown from a mere 39 cooperatives with a membership of 1,400 miners (Chiwawa, 1993) in 1985 to 46 cooperatives with a membership of over 4,000 miners in 1995. As required by the tribute agreements, chromite cooperatives have maintained supply contracts with both Anglo American Corporation and Union Carbide, the claim holders. Some analysts view this arrangement as anti-development or exploitative as cooperators are denied the opportunity to seek alternative markets for their output.

Cooperative tin and tantalite mining also began at the height of the depressed base metals market in 1985. Following the collapse of the international tin market in 1985, the tin mining company, Kamativi Tin Mines decided to downsize operations through retrenchments. Through the ZMDC, the government encouraged miners to organise themselves into cooperatives to continue with mining operations with some government assistance. Compared with their chromite counterparts, tin and tantalite cooperatives continue to face problems due the depressed tin prices. The result has been a reduction in the number of tin cooperatives. In fact, recent attempts have centred on trying to revitalise the sector through injection of capital from international mining companies. Other minerals such as gold, and copper are also mined on a co-operative basis in Zimbabwe.

Informal mines which include the lone operators and family ventures are scattered all over the country. They constitute the largest proportion of small scale mines in Zimbabwe, especially if alluvial gold panners on Zimbabwe's major rivers are included. In addition to over 35,000 formal small scale miners, there is an estimated 150,000 informal miners engaged in alluvial gold panning along the Mazowe, Insiza, Tokwe, Shashe, Tuli, Umzingwane and Runde rivers. It is estimated that approximately 4,600km of rivers are being panned for gold in Zimbabwe. However, no reliable gold production statistics are available from panning operations since most of the output is lost to the underground economy or parallel market. Analysts estimate that over ten tonnes of gold are lost to the parallel market every year prejudicing the country of revenue in excess of several billions of dollars. Holloway describes informal small scale miners as islands of prosperity in a sea of poverty.

However, both formal and informal small scale mines have continued to expand in scope and scale. In fact, prior to independence in 1980, there were very few formal small scale mines. Those that existed were seasonal, produced gold and were owned by individual white farmers. However, the post 1980 era has seen the mushrooming of small scale mines concentrated in the search for gold, copper, tin, diamonds, other gemstones and chromite. Maponga (1993) using 1988 data classified 10 per cent of Zimbabwe's mines as small employing between one to fifty workers with the large mines employing the remaining 90 per cent. In the same study, 20 per cent of the country's gold mines were classified as small, producing between zero and fifty thousand tonnes of ore per year. In the case of chromite producers, 75 per cent of the mines fell in the small scale category (Maponga, 1993).

In the last ten years gold panning has been on the increase in Zimbabwe being fuelled by a myriad of factors among which are; tough economic conditions leading to worker retrenchment, closure of some large scale mines (mostly base metal operations), poor agricultural yields due to drought, a new 'gold rush' phenomenon and the general lack of employment opportunities for secondary school graduates. In some cases, panning is noted to be on the increase during the dry season when peasant farmers leave their land for panning along the rivers. Gold panning provides an alternative economic base to the peasant community especially with the erratic rain pattern which characterises Zimbabwe today and as such is taken seriously by panners. In fact, panners on the Mazowe river valley confirmed that panning income lubricates the agricultural sector by providing money for fertiliser and seed.

The fact is small scale mining remains an important sector in Zimbabwe's mineral production. The sector's contribution to total mineral production, export earnings, employment generation, provision of inputs into the local and regional economy and other socio-economic parameters continues to expand. In effect, analysts agree that the multiplier effects of small scale mining are persuasive and significant. Ghose (1993) points out that, given an enabling environment, the small scale mining sector can make immense contribution to the economic development of poverty stricken communities in the developing world. He cites examples of chromite mining in Zimbabwe (the subject of this paper) as an illustration of how prudent government policies can help the growth of small scale operations and hence economic prosperity. The point is, small scale mining in Zimbabwe has prospered under the veil of favourable institutional support structures. In the forefront of small mines promotion is the Small Scale Miners Association which offers invaluable assistance to would-be miners in both prospecting for minerals and licensing of claims. The efforts of the Government through the Zimbabwe Mining Development Corporation's (ZMDC) 'mine development fund' a form of revolving fund and its plant hire scheme has helped sustain the chromite and tin mining sectors. In fact, for the period 1985–1992, a total of 46 chromite mining cooperatives have been formed, a direct result of this Government supported initiative. Another important institution in small scale mining promotion in Zimbabwe is the Shamva Mining Centre set-up with the aid of Intermediate Technology Development Group (ITDG) in the heart of the country's small scale mining area — east of the capital, Harare. The centre offers crushing facilities for the recovery of free gold.

After this general introduction, the paper now moves to review the socio-economic and environmental impacts of the two small scale mining sectors, that is, chromite mining and gold panning.

## Small Scale Mining and the Environment

The major problems associated with small scale mining are to do with the sector's relationship with the environment and the adverse effects on the lives of the miners. It is argued that compared with large scale mining, small scale mining has a greater propensity to affect the environment in an adverse manner and hence the sector has been accused of being the worst contaminator of the natural environment by some analysts — see Noetstaller (1993). However, Ghose (1993) argues that the magnitude of environmental wear, volume of residuals generated and the total impacts from this sector are much smaller and more easily controlled than with large scale activities. The question is whether the small scale mining sector does anything about its environmental waste or not.

The major environmental problems associated with the activities of small scale miners include; destruction of vegetation, hydrological disturbances, noise and air pollution, contamination of surface and underground water, uncovered mined-out areas and workers safety. Many of the environmental problems are due to the technology used in extraction and processing, lack of mining and processing know-how, lack of financial and material resources to install mitigative measures or programmes and at times negligence by the miners.

The environmental problems from small scale mining vary from sector to sector or from mineral to mineral and are encountered during both the production and reproduction stages of the mineral supply process.

*The Chromite Mining Sector and the Environment*

Cooperatives mines are an important arm of Zimbabwe's mining sector being the sole producers of chrome in the country. In chromite mining, there is widespread employment of artisanal mining methods involving use of human energy aided by rudimentary tools. Environmental problems such as; land disturbance, deforestation, depletion of water resources, and air and noise pollution are uncommon in chromite mining areas of the Great Dyke.

## Land Disturbance

The chromite ore occurs in such a manner that both surface and underground mining methods are employed. Surface trenching, adit and winze are the common mining methods used on the Great Dyke. Surface trenching or pig rooting is used where the chromite ore out-crops on the surface and involves very little mechanical effort. After exhaustion of the orebody, the trenches are abandoned and left uncovered. Chiwawa observes that most mining trenches are abandoned when they reach a depth of ten metres as hoisting of ore from such depths becomes difficult. It is therefore common to encounter these deep trenches on the Great Dyke. Both human and animal life is endangered by these trenches. In adit mining, horizontal tunnels are constructed to the chromite seams on the hills of the Dyke. Explosives and hand held and electric drilling equipment is used to open the area. Waste rock is transported to the surface and dumped indiscriminately in the surrounding areas. Timber is used as underground support and lining of tunnels. All the mining methods result in land disturbance and problems of waste disposal. With surface mining, the overburden is stored externally and thus the Great Dyke is littered with heaps of waste rock which in certain instances has sterilised fertile agricultural land around the North Dyke area.

**Deforestation**

Destruction of vegetation occurs initially in the process of clearing workings and construction of access roads to the orebodies. Chiwawa observes that adit and winze mining require timber for tunnel support and this is sourced from the surrounding farms. In surface trenching and open pit operations, bulldozer operations result in the indiscriminate felling of trees in road construction. All these activities culminate in the dissipation of vegetation cover and loss of genetic diversity of plants and animals as natural ecosystems are disturbed. Though one might be tempted to argue that large scale operators would cut trees as well, the point here is that the co-operators do nothing or very little to restore the vegetation cover. Their counterparts in the large scale sector invest in revegetation and slope stability programmes to restore the environment to its original state.

Removal of top soil by agents of erosion becomes easy where there is little or no vegetation cover. Day to day chores of the co-operators and their families also results in vegetation destruction and scouring of the land. For instance, wood is the major source of fuel and the principal building material used. In fact, 93 per cent of the energy requirements of the co-operators and their families are satisfied through use of fuelwood (Chiwawa, 1993). This puts pressure on the naturally growing forest trees which co-operators do nothing to replenish. Abandoned mining areas are bare of any vegetation, a testimony to the devastating nature of reproductive activities of the miners and their families.

Destruction of vegetation is further exacerbated by the lack of any planned settlements in the mining areas. Settlements are everywhere and encroach into farming land which borders the Great Dyke. Another contributing factor to erosion is that miners resort to market gardening to supplement their meagre income and this accelerates erosion and contributes to river sedimentation and hence disturbance of the water courses.

**Depletion of Water Resources**

Underground chromite mining results in water disturbances causing flooding of workings which requires dewatering leading to the creation of undesirable ponds and marshes around the mining areas. Such water disturbances result is falling underground water levels and the alteration of underground water supplies. Erosion of the bare overmined forests results in tons of waste choking the river system.

## Air and Noise Pollution

Little or no chemical pollution results from co-operative chromite mining since no mineral processing is undertaken on site. However, day to day reproduction activities result in some degree of pollution mainly through the burning of fuels and use of water pump engines, compressors and bulldozers. Blasting operations result in noise and dust pollution for the inhabitants. Dust pollution affects plant life in the vicinity of the mining operations and has long term effects on the health of miners.

### Gold Panning and the Environment

The upsurge in gold panning in Zimbabwe has led to the destruction of natural habitats existing along target rivers through vegetation removal, landscape distortion, river siltation or the poisoning of the riverine environment. In order to appreciate the environmental impacts of gold panning, the results of a survey carried out on one of Zimbabwe's major panning sites are used in the following discussion. Panning is defined to include traditional sand washing and horizontal tunnelling.

This particular study covered the gold panning communities along the Mazowe River and its tributaries in the Mashonaland Central Province in North Eastern Zimbabwe — see Fig. 1. Some 18km along the Mazowe river's course and its tributaries was covered in the survey. The river dissects the Harare-Shamva greenstone belt which is made up of andesitic and dacitic metavolcanic rocks of the Bulawayan group as well as metasediments and felsic metavolcanics of the Shamvaian group. As such alluvial gold found in Mazowe River and its tributaries might derive from these Shamvaian and Bulawayan rocks. The processes of eluviation and illuviation result in gold being deposited in river bed rubble, riverbanks and with silt in the flood plains. In addition, the catchment area of the Mazowe River is underlain by basaltic greenstones of the Grahamsdale formation, which outcrop and occasionally have contact with the river system. A hill close by, Tafuna, has a diversity of greenstones and could be the source of the reef mined from the hill by gold panners. Thus, alluvial gold panning activities along the Mazowe River attest to the extensive gold deposits derived from the greenstone belts in the river's catchment area. Such a geological environment provides adequate stimuli for alluvial gold panning along the Mazowe River.

Environmental problems emanating from panning activities during both mining and processing are of immense proportions on the Mazowe valley and the adjacent farms. Most observers concur that the most serious impacts of gold panning are; inappropriate use of mercury which causes both air/water pollution, deforestation, encroachment into prime farming land, direct human poisoning, and hydrological disturbances due to sedimentation.

## Deforestation

Forests are the livelihood of panners in the Mazowe valley as wood is an important input in the gold panning communities. Forests are the sources of:

- hut (house) construction materials
- firewood,
- handles for digging equipment,
- mine support, and
- manufacture of panning dishes and prospecting branches.

## Construction Materials

Dwellings of gold panners are constructed from small twigs and straight poles and this means that trees are cut as panners move from place to place and new homes are constructed. The fact is, panners are nomadic. The number of abandoned huts along the river valley clearly indicates of the periodic shifting of homes and the potential impacts on forests. The use of straight poles of almost the same thickness results in immense forest destruction, for one hut might require the felling of some fifty trees or so to get enough straight poles. An equal amount of poles is also required for the roofing of the houses and this exerts immense pressure on timber resources. A visual inspection of the tree population around the valley, testifies to indiscriminate and heavy tree felling.

## Fuelwood Requirements

Firewood is the chief source of fuel and most of this is illegally acquired from nearby farms or reserved areas set aside by the Natural Resources Board. Under normal circumstances collection of firewood is not very destructive to the forests as only dry wood is collected for use as fuel. This is not the case on the Mazowe valley where green wood is used due to the scarcity of dry wood. As harvesting of naturally growing trees is excessively higher than natural regeneration and with not tree planting taking place, the valley is slowly becoming thinly vegetated. Figures 1 and 2 show the extent of vegetation destruction along the valley. Trees are

Fig. 1

Fig. 2

Figs. 1 and 2: Vegetation destruction along the valley

unable to regenerate naturally due to continuous cutting down by panners and inteference by domestic animals.

## Panning Dishes

To recover free gold from stream sediments, panners used panning dishes — see Figs. 3 and 4. Crushed ore is put in the wooden dish and using water and rotating the wooden dish from side to side, panners use the density of materials to separate gold from the dirt. Panning is performed everywhere, from streams to reefs where artificial water ponds are created for washing the sand. Wooden pan dishes are used by almost all (99 per cent) of the gold panners and this has serious implications for the rate of harvesting of trees. From the writer's own observation, a single panning dish requires a tree of about 40–50 cm diameter and has an average life span of three years. This means that with the numbers of people involved in panning and the extent of use of wooden dishes, there is uncontrolled tree felling for making these dishes. In fact, the making of panning dishes has become a source of revenue for some communities in the valley.

## Mining and Mine Support

The first stage of mining, exploration, results in indiscriminate cutting down of trees to open up areas. In addition, panners destroy trees as they follow gold reefs — see Fig. 5. The cranks (Fig. 6) for conveying gold ore from the pits are made of wood and so are the underground support structures which are made from the selected wood harvested from around the area. These have a very short life of less than two months and in most cases are abandoned as panners move to new sites. Further, the makeshift ropes on cranks are obtained from the fibre barks of trees and this stunts the growth of trees and in some cases, the debarked trees wilt prematurely.

Regrettably, the whole panning cycle from clearing up areas during prospecting up to processing of gold is accompanied by the destruction of vegetation.

## Land Degradation

Land degradation by gold panning is extensive. Primary land degradation includes, the direct impact of all the panning operations, mining, processing and site clearance. Visible direct effects are undercut river bank and deep trenches across dry river beds. The destruction of banks has widened and disfigured the river channels and results in loosened soil from the banks being discharged into the river channel once rain falls. Tunnels,

Fig. 3

Fig. 4

Figs. 3 and 4: washing sand for gold on the Mazowe River. Wooden dishes used.

Figure 5: Reef prospecting and vegetation destruction.

Figure 6: Hoisting gold ore.

pits, trenches, and pot holes destroy flat terrains of some areas (see Fig. 7). Some of these undeground tunnels collapse and result in land subsidence. Tunneling is commonly used to extract gold from river banks and the resultant trenches provide a conducive environment for gully erosion and the creation of huge dongas, some with channel widths and depths in excess of 10 m and 15 m respectively as observed along one of the tributaries. The final gold recovery sites are ugly dump heaps which cause mudflows into the rivers during the rainy season. The dumps and mudflows destroy the underlying vegetation and cause extensive river siltation.

The scraping of the top layer further exacerbates sheet wash erosion as the surface is left bare making a platform where water can flow at a high speed and this is worsened by the clearance of vegetation which makes the ground vulnerable to soil erosion as the rain drop bomb is not intercepted. Accelerated soil erosion results in the discharge of excessive loads of silt and loose materials into the river causing floods through blockage of channel flow.

Furthermore, deep trenches characterise most panning sites and these pose dangers to both animals and the panners themselves. For instance, at one panning site, along the dry tributaries an area of 67.5 $m^2$ had six pits some of them interconnected by tunnels (see Fig. 8).

The fact is, the operations on or next to water courses, and as such they have a disproportionately large impact on the riverine ecosystem. In actual fact some researchers have argued that sedimentation of water courses is the most significant of all environmental costs attributable to small scale mining. For instance, in Brazil's garimpos, scientists estimate that two cubic metres of sediments enter the water courses for every gram of gold produced (MacMillan, 1993, p. 157). Although sedimentation varies due to host rock conditions, mineral extraction methods, gold panning results in the deposition of tons of waste into the Mazowe River due to the amounts of loose material the process generates. This not only interferes with the reproduction of fish, which are an important dietary source of protein for the panning community, but also accelerates the rate of destruction of water reservoirs. Sedimentation has resulted in the extinction of some of Mazowe's tributaries and reduced some to mere streams as their course is punctuated by islands of sand, grass and small bushes. However, it would be unfair to attribute sedimentation to panning only, the river cuts through rich agricultural farming areas and therefore some of the sediments could derive from the farms.

The removal of top soil during mining has serious long-term effects as building soil profile takes many years during which the productive capacity of the land is lost. Regrettably, the detrimental impacts of panning have not been confined to the river beds and banks. The activity has provoked violent land conflicts in certain parts of the panning areas.

Fig. 7a: Trenches on the valley

Fig. 7b. Panned Terrain

Fig. 8. Holes on the panned areas.

## Mercury Contamination

There is widespread use of mercury on the Mazowe River valley, in fact, mercury is an integral part of mining 'technology' in gold recovery. Once ore has been won it goes through a series of poundings (see Figs. 9 and 10) into fine powder after which the gold ore sludge is alloyed with mercury into an amalgam which is then separated by heating. The process is fairly simple and has a comparatively high rate of recovery. Its first application dates back to mining in Bosinia under the Emperor Nero (54 to 68 AD) and up to the present, small scale mining has continued to employ this technique intensively — see Neisser, 1993.

The process is suitable for already liberated elementary gold that has not been encrusted, for example, fine ferrous oxides, and ranges between 20–50um and 2 mm. The lower grain size limit largely depends on; the interfacial tension of the mercury and water and the shape of the grains. In formal placer mining the already liberated gold is combined with the mercury only. For this process riffled sluices are mainly used, in which the gaps between the riffles are filled with mercury. The entire feed is sluiced through the riffles as a slurry. During this process the slurry absorbs between about 5%–30% of mercury from riffles, but subsequent devices for recovering this mercury are lacking. Large amounts of soap or similar surface active agents are added to increase the surface of beneficiation while reducing mercury loss — see Neisser, 1993. Primary ores require

Figs. 9 and 10. Crushing ore before amalgamation.

the liberation of the valuable minerals which the miner amalgamates either during crushing or in a special processing phase following milling. To amalgamate and crush simultaneously, panners use edge or chilean mills or crush manually. For separate, subsequent amalgamation; amalgamation barrels, amalgamating copper plates and manual amalgamation in gold pans are used. On average, losses from the metallic mercury from the sorting and amalgamation equipment comprise about 40%–50% of total losses — Cleary, 1994. The amalgam-mercury mixture thus obtained is separated into the highly viscous mercury and liquid mercury by squeezing it through leather or cloth or old blanket.

Panners along the Mazowe River apply much more rudimentary methods of amalgamation than formal miners. To recover gold, the amalgam bat with about 50%–60% by weight of mercury and 40%–50% gold wrapped in paper in an open crucible is heated on an open fire at temperatures of 350–600 °C. This takes place under the cover of darkness and indoors. People who inhale mercury from the burning process absorb the metal in an organic form through the mucous membranes of the nose or lungs. However, MacMillan points out that inorganic mercury is not much of a problem as it can be flushed out of the body and discharged through urine.

Problems arise with the disposal of used mercury and air pollution from contamination by fumes from heating of the amalgam. In door heating of the amalgam further exacerbates the problems of air pollution and direct poisoning of panners. Both legal and illegal panners prefer indoor heating for different reasons though, the former for energy conservation and the latter to avoid the wrath of the law. The deposition of unused mercury is likely to give rise to raised mercury levels in soils, with subsequent uptake by plants and raised mercury levels in sediments and waterways. The used mercury finds its way into the river system and thus poisons domestic animals and aquatic life which depend on the river water for life.

The contamination of rivers with elemental mercury leads to increased uptake and absorption of methyl mercury in fish eating communities of the Mazowe valley. In the course of movement of mercury along the food chain, mercury is transformed from inorganic to an extremely toxic organic state, called methyl-mercury through the process of methylation. In the end human beings consume highly concentrated mercury in fish. In fact, fish is the most common relish for the panning community. Though no mercury level measurements have been undertaken in the panning community, all indications are that the people have high mercury levels in their blood.

Some gold panners falsely believed mercury is harmless since it doesn't stay in one's stomach if swallowed. In addition, excess mercury is handled by bare hands and also roasted indoors without any protective clothing.

The sequence of burning has to be borne in mind when assessing the likely environmental impact of mercury vapour released as a by-product, since the type of burning which takes place in the location where the gold is extracted and those that occur as it moves along the trading chain are significantly different. Thus, there is inhalation of mercury fumes as no masks are used during the burning process. Inhalation of the fumes also affects other people within the vicinity of the burning sites as air is a public good. Studies indicate that organisms at the higher trophic levels of the food chain tend to accumulate much more concentrated mercury levels and this implies that humans are subjected to this highly toxic substance when they consume fish.

## Hydrological Disturbances

There is evidence of extensive disturbance of the hydrological system of the Mazowe River and its tributaries in terms of the flow of water and the catchment area. The processes of deforestation and land degradation discussed above subsequently choke the river channels through siltation. Gold panning also lowers the water table as the pits dug along the river banks and the flood plains are lower than the natural water table in most cases and as the water is pumped out with the ore the surrounding water table lowered even further. To compound the problem, tunnels and pits collect water thus reducing the catchment area of rivers. The fact is, mining alluvial gold provokes changes to the aquatic ecosystem. All mining techniques used disturb river and stream sediments, increase siltation rates and lead to drastic alteration of aquatic life through an increase in the levels of suspended material which adversely affects fish. Fish is one of the principal dietary components of the panning communities and as such a reduction in fish population due to siltation of channels when juxtaposed with uncontrolled fishing results in reduction in food in the upper trophic levels in the long run.

Siltation and lowering of the water table have resulted in the drying-up of some portions of the river and its tributaries. Some tributaries are now extinct while others have retreated as panners have cut more trenches upstream. The late fluvial disturbances of panning and sedimentation have disfigured the river channels.

## Working Conditions, Health and Sanitation of Miners

Communities, albeit temporary, have sprung up along the river valley and with them have emerged serious health problems. Sanitary facilities and water sources are substandard. Shallow pit latrines (less than five metres deep in most cases) are used by 98 per cent of the gold panners and these are allover the place. Most panners acknowledged to using the open bush system during the day. No piped water is available at the panning sites and 32 per cent of the miners drink untreated water from the disused mine pits, 27 per cent use river water and 21 per cent of the miners drink from open wells, exposing themselves to diseases. Due to these conditions, dysentery and cholera outbreaks are common in the valley, a fact confirmed by health officials in the area. Statistics from the local hospital indicated that in 1994 alone, a cholera outbreak killed more than 10 gold panners at a squatter settlement in the Mazowe River valley.

The quality of housing is poor and would be inhabitable under normal circumstances — see Fig. 11. In most instances three to four people share a tiny room which would under humane circumstances accommodate only one person. The houses are overcrowded, small, neglected and holed all over giving minimum protection from the weather, mosquito and other insect bites.

Another disturbing feature along the Mazowe valley is the dangerous working environment (Figs. 12 and 13). Gold panners work with no protective clothing whatsoever, not even hard hats or boots. In fact, some underground panners were observed to be working with very little clothing on. Accidents due to the collapse of underground workings are frequent and high fatalities have been recorded. The major causes of accidents are unplanned tunnelling, cuts by stones and equipment, collapsing of tunnels and pits, falling in disused mine shafts and dangers of being hit by falling stones. Unfortunately, the gold panners are not equipped with any first-aid skills to help casualties and the panning sites are at least 10 km from the nearest hospital giving seriously injured miners slim chances of making it to the nearest hospital. Tunnels and pits are poorly ventilated and lit with candles and at times smoky makeshift paraffin lamps subjecting gold panners to respiratory problems.

*The Government and Environmental Management of Mining Areas: The Issues*

Given that there are several pieces of legislation to monitor and control environmental damage from all mining activities, the question is why these problems persist. Government efforts to solve these problems have been through legislative instruments. The legal instruments that come to mind are; the Mines and Minerals Act, Mining and Safety Regulations, Hazardous Substances Act, the Mining (Alluvial Gold) (Public Streams) Regulations and many others directly and indirectly related to mining. It is clear that the problem is not about lack of regulations and standards but weaknesses in implementation and adherence to these standards.

The use of a command and control approach to manage the adverse environmental effects of small scale mining has not met with a lot of success. This can be attributed to the nature of the sector, very informal.

Fig. 11. Typical accommodation in the panning villages.

Fig. 12. Underground tunnel.

However, most writers concur that the ineffectiveness of legislation could be explained by any of the following factors;
- lack of awareness of standards,
- monitoring weakness,
- lack of mitigative resources,
- unclear legislation, and
- lack of understanding of long term benefits of good environmental sterwardship.

## Regulation Awareness

Regulations are effective if they are known and understood by their subjects. Small scale miners pronounce ignorance about certain environmental standards. In the case of panning, though there was massive violation regulations were known (and presumably understood) by over 56.5 per cent of the panners they were contravened by most panners (95 per cent). Panners challenged some of the tenets of the legislation as impractical and believed they were meant to stamp out alluvial gold panning. For instance, panners argued that by restricting panning operations to the river bed and/or at least three metres from the lowest point of naturally defined river bank, they almost made panning impossible during the rainy season because of the high water levels. This, they argued, relegates panning to a seasonal occupation which is not practical for the Mazowe River community which regards panning as a full-time activity. The maximum pit depth of one and a half metres specified by the regulations was unacceptable to most of the gold panners on grounds that in most cases gold deposits occur on the bedrock and are/were rarely found within this prescribed limit. Panning thirty metres away from the river banks as stipulated by the regulations was uneconomic as it increased the costs of fetching water. Regulations almost defined the rivers to be panned and this was regarded by panners as being unrealistic.

## Monitoring Problems

Manpower and financial resources are the major handicaps to monitoring compliance in environmental standards from the small scale mining sector. For instance, the panning regulations of 1991 empower district councils with the responsibility to ensure rehabilitation of mined-out areas. Neither manpower training nor financial resources are provided for these functions and hence its not suprising that nothing or very little is done by the councils. The government's Mines Inspectorate within the Ministry of Mines faces the same dilemma as mines inspections cannot be undertaken periodically as required by law due to lack of finance and manpower for this purpose. Hence environmental mismanagement continues unabated as the Ministry only responds to crisis areas.

## Lack of Mitigative Resources

Rehabilitation has got several handicaps to it. Mine rehabilitaion is a low priority for the small scale miners. As discussed elsewhere in this paper, small scale mining is a subsistence activity generating little or no surplus for 'luxury' purposes such as rehabilitation. To the miners there is not time, energy and motivation to backfill a mined-out trench, the focus is on the next panning site. This is clearly illustrated by the attitude of panners on the Mazowe valley. Though over 50 per cent of the panners were aware of the rehabilitation regulations, no attempts to rehabilitate the panned-out areas were observed for various reasons. Some panners believed that the pits and the tunnels would backfill by soil erosion during the rainy season and hence saw no point in backfilling and wasting energy. Other panners argued that the price of gold offered on the parallel market was too low to give them the time and incentive to rehabilitate the panned-out areas. There was also a general feeling that the rehabilitation burden should be borne by the claim owners as some panners were tributors. Even though most gold panners were aware that panning damaged the environment, only a small proportion (25 per cent) was willing to contribute towards a rehabilitation.

## Unclear and Fragmented Legislation

Some of the current problems in the effectiveness of mining legislation can be attributed to unclear pieces of legislation. This is further exacerbated by the fact that for the mining sector there is not one piece of legislation dealing with environmental management, one has to go through at least seven or eight instruments before they have a complete environmental document. This causes partial satisfaction of certain requirements of the law. Such important legislation should be in simple language and handled by one controlling authority with all the expertise.

## Discussion

The preceding discussion touched briefly on the benefits of small scale mining but dwelt extensively on the negative side of the activity, lack of proper environmental management practices. Solution of these problems requires an understanding of the causes and designing appropriate measures to rectify the situation. Textbook solutions are limited in solving small scale mining problems as these are site specific with very little generalisation in most cases.

It is undoubtable that small scale mining be it gold panning or chromite mining causes serious environmental problems during both production and reproduction processes. Small scale mining activities can no longer be regarded or dismissed as ad hoc activities; they have proven to be part of the development process of many countries and Zimbabwe is no exception. This study has brought into focus the problems associated with gold panning and chromite mining activities in Zimbabwe and attempts by government to minimise these adverse effects.

Given the general problems highlighted in this paper, the following recommendations are proposed as possible remedies.

## Legislation Clarity and Manpower Training

Strengthening of existing alluvial panning and chromite mining legislation by either provision of additional well-trained (environmental training) manpower to monitor environmental management periodically. The point here is that command and control environmental management requires a strict monitoring system to ensure compliance. Any existing conflicts within legislation should be eliminated and important pieces translated into simple and understandable language for the poorly educated panners and miners. This also includes bringing under one umbrella ministry, issues related to environmental management. Authorities charged with monitoring should be provided with adequate financial and manpower resources.

## Educational Campaigns

Educational campaigns should be run through both the electronic and print media and workshops to sensitise the communities about the benefit of rehabilitation and good environmental stewardship. Pilot environmental management schemes should be run in the mining communities to illustrate the benefits of multiple land use. In fact, before a prospective claim holder is issued with a certificate, or before registration is approved, the council should educate the miners on the environmental problems and the possible ameliorating methods that could be used. Alternatively, rehabilitation plans for panned-out areas should be provided before registration is permitted or approved as a requirement for pre-mining EIA. This would only be feasible if the district councils had the financial resources to insist on such plans and were able to stamp-out those panners who chose to become illegal.

## Marketing of Minerals Output

The buying of gold produced by the panners should be decentralised to within reasonable distances of the panning sites. This would tap the gold which is currently lost to the parallel market and ensure that miners obtain competitive prices. Payment of competitive prices would ensure that panners are able to finance rehabilitation programmes, since under the current system the panners link their inability to rehabilitate river banks to the low prices paid for their output; as a result they are always in a rush to pan for more gold and with it more income. The same holds for chromite mines, the marketing agreements with Zimasco and Zimalloys are skewed in favour of the mining companies as they control what price they pay to the miners. Poverty is the cause of environmental damage, thus elimination of poverty through increasing income will solve some of the environmental impacts from small scale mining. Chiwawa notes that of the chromite mining families who had access to electricity and possesed electric stoves, a mere 14 per cent used these stoves for cooking, preferring instead to used 'freely' detained firewood due to the high cost of electric energy. In this case raising income levels through paying competitive prices will partially reduce cutting down of trees for fuelwood purposes.

## Introduction of New Technology

The amalgamation process poses great health risks to the panners especially during the heating of the amalgam as this is done mostly indoors. Portable retorts, which are much cheaper should be made available to the panners through the Government's hire scheme. Alternatively, the government should actively support investment into clean technology through fiscal means such as tax concessions. Its worth mentioning that new technology should be matched and supported by appropriate promotional policies, legislation and technical services.

Poor recovery technology in panning is wasteful in that only free gold is recovered through either amalgamation or the James Table (see Fig. 14 for a typical James Table) and material locked up in pyrite is thrown away as waste in tailings. At one panning site, tailings analysed for gold showed levels of over 5g/t, which is wasteful. The important thing is that panners should be introduced to better and more efficient recovery technology otherwise they will continue to waste the country's important resource.

## FINAL THOUGHTS

Migration to gold panning and chromite mining sites cannot be separated from the impacts on the agrarian economy. In the survey, more than 70 per cent of the panners had experience in agriculture which implies that the workforce was drawn from active agricultural world. This could have serious implications as a bandwagon effect could result in agricultural labour moving to the rivers may be agriculture should be self-supporting.

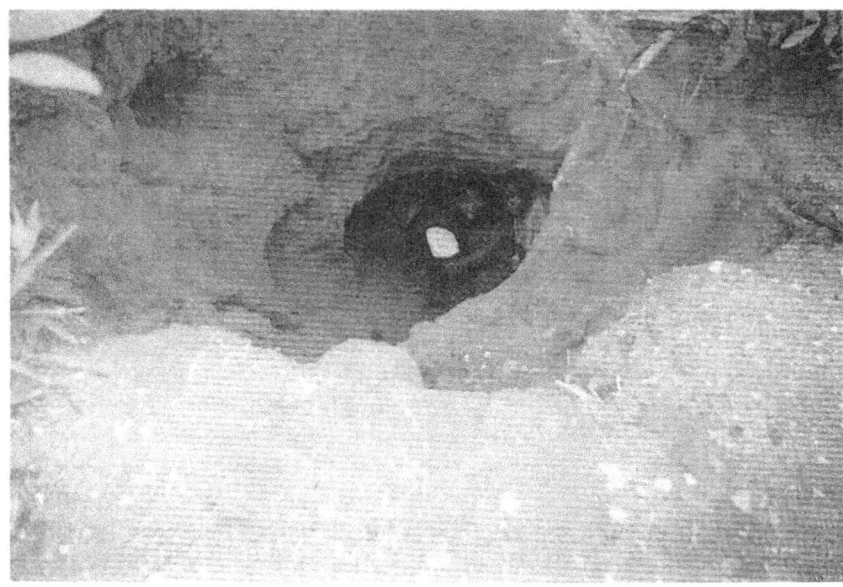

Fig. 13. Underground mining of gold.

Fig. 14. James table.

Much of the case against small scale mining hinges around its adverse effects on the environment but what must be understood is that environmental degradation is a manifestation of poverty and that reduction in poverty is the solution to this problem. In this vein, small scale mining as evidenced by field studies is contributing to the alleviation of poverty which will ultimately result in reduced environmental problems as miners can use their acquired wealth and resources to do something about their livelihood, the environment. Afforestation and reclamation of mined-out areas can only be considered a priority after basic needs have been satisfied. The successes achieved in tourism, agriculture and other industries where there has been considerable small scale investments by local entrepreneurs is sufficient evidence that industrial growth and thus economic prosperity can be fostered through the promotion of small scale industries. An injection of capital into small scale mining can go a long way into alleviating the adverse environmental effects of the sector. A level playing field in product marketing would significantly improve the ability of miners to rehabilitate mining areas. Education campaigns on environmental management and government assistance small scale miners can minimise environmental damage from this sector. The removal of impediments to the development of small scale can go a long way in promoting better environmental management.

**REFERENCES**

Shoko, M. (1993). *Report from the First National Workshop on Mines and the Environment*, Department of Natural Resources and Chamber of Mines, Gweru, Zimbabwe, November.

Cleary, D. and Thornton, I. (1994). 'The Environmental Impact of Gold Mining in the Brazilian Amazon, *in Hester, R.E. and Harrison, R.M. (eds). Mining and its Environmental Impact.* Cambridge; Royal Society of Chemistry (Issues in environmental Science and Technology)

Anderson, F.R., Kneese, A.V., Taylor, S. and Stevenson, R.B. (1977). *Environmental Improvement through Economic Incentives: A resources for the future book*, John Hopkins University Press, London.

Henley, D.C. (1989). *A preliminary Analysis of Zimbabwe's Natural Resource Legislation and a Proposal for an integrated Approach to Resource and Environmental Management in Development: A Draft Report to the Canadian International Development Agency and the Ministry of Natural Resources and Tourism*, Harare, Zimbabwe.

Whitlow, J.R. (1990). 'Mining and its Environmental Impacts in Zimbabwe.'*Geographical Journal of Zimbabwe*, No. 21, pp. 50-80.

Maponga, O. (1996). Small Scale mining and the Environment in Zimbabwe: The case of alluvial gold panning. *Paper prepared for presentation at SWEMP96, Cagliari, Italy,* September 1996 (Forthcoming).

Maponga, O. (1995). *Small Scale mining in Zimbabwe: An agent for destruction or development.* A paper prepared and presented at a UNESCO Regional Conference on Minerals Development, Entebbe, Uganda, October 1995.

Maponga, O. (1993). *Small Scale Mining Operations in Zimbabwe.* International Development Research Centre Publishing (IDRC), Ottawa, Canada, 23pp.

Maponga, O. and Mutemererwa A. (1995). *Management of Natural Resources and the*

*Environment in Zimbabwe: the case of Gold.* UNCTAD\COM\45. February.

Neisser, W.E. (1993). *Problems associated with the use of mercury by small scale gold miners in developing countries.* Unpublished Master of Engineering Thesis, McGill University, Montreal, Canada, 104 pp.

van Blerck, M.C. (1994). 'Mining and the Environment: Tax Incentives encourage orderly planning'. *Journal of the South African Institute of Mining and Metallurgy,* June, pp. 129-132.

Davidson, J. (1993). The Transformation and successful development of small-scale mining enterprises in developing countries, *Natural Resources Forum,* 17(4), pp. 315-326

Department of Mines, Botswana (1984). Small Scale Mining in Botswana, 30 pp.

Newman, C.J. (1989). Small workings in Zimbabwe: recklessness or economic risk. *Extracts from Small Mines Economics and Developments Conference, Imperial Collegel London.*

Noetstaller, R. (1993). Small Scale Mining: Practices, Policies, Perspectives, *in Ghose A.K. (Ed) Small Scale Mining: A Global Overview,* Oxford, New Delhi, pp. 3-10.

Ghose, A.K. (1993). New Configuration for Small Scale Mining for Developing Countries, *in Ghose A.K. (Ed). Small Scale Mining: A Global Overview,* Oxford, New Delhi, pp. 29-42.

Hollaway, J. (1991). Role of Small Scale Mining in Africa: Building on the informal sector, *in African Mining'91,* IMM, London.

Hollaway, J. (1986). Small Scale mining Sector in Africa: Restructuring for profitability. *Natural Resources Forum,* 10(3), pp. 293-297.

Wright, E.A. (1993). Environmental cost considerations in planning and operating small mines. *Chamber of Mines,* 35(3), pp. 30-33.

Killick, T. (1974). The developmental impact of mining activities in Sierra Leone, *in Scott, R. Pearson and John Cownie, Commodity Exports and African Development,* Lexington Books, Lexington, pp. 217-235.

Stewart, D.F. (1989). Large-scale vs small scale mining: meeting the needs of developing countries, *Natural Resources Forum,* 13(1), pp. 44-52.

Chiwawa, H. (1993). *The political Economy of Small Scale Mining in the SADC. Institute of Development Studies,* University of Zimbabwe, 32 pp.

Chiwawa, H. (1993). *The Environmental Impact of Cooperative Mining in Zimbabwe.* A paper prepared and presented at the Organisation for Social Science Research in Eastern and Southern Africa, 4th Congress on The Global Nature of the Environmental Crisis and its interrelationship with Development: Africa's Plight, Debre Zeit, Ethiopia, August 9-12.

Mutagwada, W. and Hangi, A. (1995). Environmentally sustainable artisanal gold mining in the Lake Victoria Region, Tanzania., *in African Mining '95, IMM,* London, pp.423-431

MacMillan, G. (1995). *At the end of the Rainbow: Gold, Land and People in the Amazon,* Earthscan, London

# SMALL AND MEDIUM SCALE MINING IN INDONESIA, HISTORY AND CURRENT ACTIVITIES - 1996

*Antony H. Osman*
Managing Director, Minconindo Pty Ltd., Indonesia

**INTRODUCTION**

Indonesia is an independent and sovereign nation which achieved its independence from the earlier Dutch colonial administration on 17 August 1945. It is a rapidly modernizing and developing country with a history stretching back to the dawn of time.

Geographically the country is an archipelago consisting of 13667 islands of which some 6000 are inhabited. It consists of three main groups of islands viz the Greater Sundas consisting of the islands of Sumatra, Java, Kalimantan and Sulawesi, the Lesser Sundas or Nusa Tenggara group consisting of the islands stretching eastwards from Bali to Timor, and the Moluccas i.e. those islands to the north of the Lesser Sundas and to the east of Sulawesi.

Throughout the centuries the islands of Indonesia, lying between the Indian and Pacific oceans, have absorbed a great number of ethnic and cultural influences. Many of those influences, most notably those associated with Hinduism and Islam, have stamped themselves indelibly on the social, cultural, economic and political processes which make up the way of life of the Indonesian nation today.

The population of Indonesia is approximately 195 million, making it the fourth most populous country in the world, and increasing at 1.8% per year. On the western end of the island of Java is located the capital city of Jakarta (founded in 1527 AD) which has a population of around 10 million. Java has 60% of Indonesia's population, is one of the world's most densely inhabited regions, and is the centre of political and economic power in the country.

The country has some 300 separate ethnic groups each with its own language and customs. The national language is Bahasa Indonesia ("the Language of Indonesia") which has developed from the Malay language with strong influence from the regional languages. The linguistic content of the national language itself reflects the strong cultural heritage derived from the Indian sub)continent, and other foreign countries, with words derived from Sanskrit, as well as Arabic, English, Dutch, Portuguese, Japanese etc.

In excess of 90% of Indonesians are Muslims with the balance being composed of Christians, Hindus (mostly found on the island of Bali), Buddhist and others.

Administratively Indonesia is divided into 27 provinces (first level areas) under a Governor and each province is further divided into Kabupatens (second level areas) under a Bupati (Regent). Further each Kabupaten is sub-divided into Kecamatans (the third and smallest level administrative area) headed by a Camat. Each Kecamatan will consist of a number of villages/Desas/Kampongs.

Indonesia is richly endowed with mineral resources apart from oil and gas, in particular gold and silver on Kalimantan, Sumatra, Java and Irian Jaya (including the copper/gold deposits at the Freeport mines); diamonds on Kalimantan; coal on Kalimantan, Sumatra, Java and Suluwesi; Tin on Belitung, Bangka and Sinkap; Nickel on Sulewesi; industrial minerals throughout the archipelago. Currently, Indonesia is undergoing a boom in exploration for mostly gold, coal and diamonds, and the small scale mining sector often comes into conflict with the larger better regulated mining activities.

## THE INDONESIAN MINING INDUSTRY

Development of the formal mining industry at all levels is strongly supported by the Government.

To put the Indonesian mining industry into perspective the main mineral production figures for 1995 are as follows.

### Gold (Silver) kgms

| Pt Aneka Tambang * | Pt Freeport Indonesia ** | CoW Coys *** | National Private | Total |
|---|---|---|---|---|
| 1812 | 42474 | 18522 | 10 | 62818 |
| (17284) | (82240) | (165689) | (9) | (265212) |

   * Government company
  ** Gold and silver in copper concentrates
 *** Gold and silver produced by Contract of Work companies (gold and silver are commonly intermixed).

Unofficial estimates put People's Mining/Pertambangan Rakyat (see later) activities at accounting for 25% of the country's gold production. This non-formal mining is believed to involve now around half a million individuals who in total produce some 15 tonnes of gold. These figures are not included in the above.

## Coal—'000 tonnes

| Govt. PTBA | Coal contractors | National Private | Kud/ Co-ops | Total |
|---|---|---|---|---|
| 7953 | 29576 | 4222 | 229 | 41980 |

Notes:
a. Government figures relate to two production areas in Sumatra.
b. Nine coal contractor companies (1 in Sumatra, 8 in Kalimantan)
c. Nine National Private (KP) producing companies.
d. Ten registered KUD cooperatives.
e. Figures do not include unregistered/ non-formal/illegal coal mining which may exceed in total 500 000 tpa

Indonesia is the world's third largest steam coal exporter after Australia and South Africa.

## Tin — tonnes

| Government PT TAMBANG | Contractor PT KOBA Tin | Non Formal | Total |
|---|---|---|---|
| 31249 | 129 | N/A | 38378 |

## Nickel

| Govt. Sector PT ANEKA TAMBANG ore tonnes | Private Sector PT INCO Ni in matte tonnes |
|---|---|
| 2513394 | 45500 |

## Copper Concentrate

| Tonnes Production | Average Concentrate | | |
|---|---|---|---|
| | Cu % | Au g/t | Ag g/t |
| 1516611 | 30.37 | 27.95 | 54.28 |

**Note** : Production is only from the Freeport mine in Irian Jaya.

**Bauxite—tonnes**

Government
PT ANEKA TAMBANG
899 035

**Iron sands tonnes**

Government
PT ANEKA TAMBANG
348371

## APPLICABLE MINING LAW

The applicable law in Indonesia devolves from European Roman-Dutch "civil law" codes.

The legal basis of mining in Indonesia today is enshrined in the Constitution of 1945 and the landmark laws of the New Order Government of President Soeharto viz. The Basic Mining Act, Law No. 11 of 1967 and associated regulations, and The Foreign Investment Act, Law No. 1 of 1967.

Small scale people's mining or Pertambangan Rakyat is defined in Law No. 11/1967 as follows:

"All mining activities practiced by local people in a simple and traditional way, using simple equipment to support their needs of daily life".

In that mining law it states that the Minister of Mines and Energy holds the authority to decide which deposits are to be allocated under Pertambangan Rakyat—in effect it is all minerals. The law also states that the deposits under consideration should be ones that are too small to be exploited profitably by large scale operations. Pertambangan Rakyat is to accommodate the exploitation of minerals by the local people in their participation in National Development in the mining sector under the guidance of the Government.

Law No. 12 of 1967, the Law of Cooperatives, provides for the establishment of Koperasi Unit Desa (KUD) or Village Cooperative Units which can be established in every Kecamatan in the country. The KUD serves as the smallest economic unit for the people living in the villages. Each KUD has its own membership area, at least one KUD for each Kecamatan. The economic activities permitted for KUD's are varied and include mining.

Besides, the above-mentioned laws there are a number of decrees, circular letters and instructions, issued by the Minister of Mines and Energy and the Director General of General Mining regulating various aspects of a technical nature and administrative procedures for submission of applications for mining authorisation, payment of deadrents, fees and royalties.

The involvement of the foreign sector in mining in Indonesia is confined to certain specified minerals in medium but preferably large scale operations under specified conditions. In practice a foreign company or a joint (national-foreign) venture company can undertake mining in Indonesia as a Contractor working directly for the Government or as a Contractor operating for a state enterprise. Foreign involvement is not permitted in small scale mining and foreign Contractors are not permitted to operate on Java. Foreign companies do however enter into financial, management or technical assistance agreements with national private companies in the medium scale sector of the mining industry throughout the country. Such agreements should be approved by the Department of Mines and Energy.

Minerals are classified into three different groups (Government Regulation No 27 of 1980 viz.

Group "A" — Strategic Minerals
e.g. Oil & gas, coal, radioactive minerals, tin, bitumen, nickel, cobalt etc.
Group "B" — Vital Minerals
e.g. Gold, platinum, silver, copper, lead, zinc, mercury, diamond, bauxite, rare earths etc.
Group "C" — Other Minerals
Minerals of lesser importance, non-metallic minerals such as limestone, dolomite, marble, granite, kaolin, various clays, sand and gravel, asbestos, talc, semi-precious stones etc.

Under the 1945 Constitution and the Mining Law mineral resources are national property and will be utilized by the State for the maximum welfare of the people. Thus the State has the exclusive mining right.

Other Indonesian parties may conduct mining only after obtaining a Mining Authorisation (Kuasa Pertambangan/KP) for Group "A" and "B" minerals, issued by the Department of Mines and Energy on behalf of the Government, or a Regional Mining Permit (Surat Izin Pertambangan Daerah/ SIPD, for Group "C" minerals, issued by the relevant Provincial Government.

KP's and SIPD's can only be granted to government agencies, state enterprises, national private corporations, Cooperatives—Koperasi Unit Desa/KUD's, or individuals. To undertake the complete range of exploration and exploitation activities, a series of KP's or SIPD's are needed viz "General Survey","Exploration", "Exploitation/mining", "Processing & refining", "Transportation & sale".

Strictly speaking, the mining of Group "A"-strategic minerals can only be undertaken by a government agency appointed by the Minister or a state enterprise. However, for economic and practical reasons, or in the interests of development in the region concerned, the Minister may designate certain, limited extent, deposits of such strategic minerals, to be mined by private national companies or cooperatives. Even small deposits of strategic minerals such as coal can be extracted by the local people using simple methods provided a "Mining Authorisation for Peoples Mining/ KP Pertambangan Rakyat" has been applied for and granted. The mining for Group "B" and "C" minerals can be carried out by Indonesian organisations or individuals who are qualified to carry out the necessary work.

## TYPES OF MINING IN INDONESIA

Mining activity in Indonesia can be categorised into a number of levels and these are:

### Medium to Large Scale Mining

State concerns (e.g. PT Tambang Batubara Bukit Asam/PTBA—the state coal mining company), Foreign Investment (PMA) companies acting as contractors to the Government (e.g. PT Freeport Indonesia—copper/gold mining) under a minerals Contract of Work (CoW) or, in the case of coal, a Coal Cooperation Contract (CCC) or the proposed Coal Contract of Work (CCoW).

Foreign interests in the mining industry basically can only operate as a locally incorporated limited liability (PT) PMA entity.

Currently, copper/gold, gold, coal, tin, nickel, bauxite, kaolin, iron sands, facing stone (e.g. granite) fall under this category.

For coal upper limit of initial exploration areas has been reduced to 100000 ha where, ultimately production is likely to be 1 mtpa+ utilizing mineable reserves of, say, 30+ million tonnes. Indonesia's largest coal company produces in excess of 10 mtpa.

### Small to Medium Scale Mining

Mining by national private companies under KP State mining companies.

Currently, Coal (non-contractor), gold, tin, kaolin, construction aggregate, sand and gravel, dimension stone, facing stone, (granite), industrial minerals.

For coal, exploration areas up to 10000 ha where, ultimately production is likely to be in the range of 200000 to 500000+ tpa utilizing mineable reserves of say less than 10 million tonnes.

## Small Scale/Village level/KUD Mining

Village level cooperatives
Currently, coal, gold, diamonds, tin, kaolin, construction aggregate, sand and gravel, industrial minerals, any mineral which has value for the local people.

For coal, exploitation areas of 10-100 ha where, ultimately production is likely to be 5000-10 000 tpa, exceptionally up to 200 000 tpa utilizing mineable reserves of, say, less than 5 million tonnes.

## Non-Formal Small Scale/Peoples Level/Pertambangan Rakyat/ Traditional/Artisanal Mining.

Currently, coal, gold, diamond, tin

## Non-Formal/Illegal Mining

Currently, coal, gold, diamond, tin, aggregate, sand and gravel, any mineral of value. Increasingly illegal mining activities, under the guise of People's Mining, are presenting serious problems for local communiities, the government, and established local and international mining companies, upon whose ground the activities take place.

## PSK—THE SMALL SCALE MINING CONCEPT

The Department of Mines and Energy have introduced the concept of PSK/Pertambangan Skala Kecil or the small scale mining concept. This concept is derived from two ideas viz 1. to assist local people who wish to become involved in the mining sector and 2. to control the illegal mining that is taking place in many provinces. The 6th five-year development program (Repilita VI) states that any mining projects involving local people should be conducted in an integrated form of small scale mining and performed within the cooperative system.

The PSK concept still requires more input from concerned authorities to result in maximum benefit to small scale miners.

## SMALL/MEDIUM SCALE MINES

### Gold

The mining of gold and silver (which metals generally occur inter-mixed together) has been carried out for centuries in Indonesia using small scale and traditional methods. In ancient times extensive underground and alluvial mining was carried out, according to the records, by "Hindu Immigrants" and the indigenous population in Sumatra and north Suluwesi. Chinese immigrants mined alluvial deposits in Kalimantan as early as the 4th century AD

Towards the end of last century there was a gold rush in exploration and mining activities but it was shortlived. Modern mining can be said to have begun in 1899 with the opening of the Lebong Donok mines at Benkulu in south Sumatra. Early in the 20th century other mines were brought into production. The total gold production from 1896 to 1941 amounted to 130 tonnes. Prior to the Pacific War there were two dredging operations in Sumatra and a few mechanized alluvial mines in Kalimantan. Prior to the 1980's some exploration for alluvial gold had taken place in Sumatra and then in the eighties the focus of interest shifted to Kalimantan. Generally speaking up to the mid-eighties gold exploration progressed at a slow rate but then rapidly changed with great interest being shown in obtaining ground by foreign companies and local entrepreneurs. The excitement again was short-lived and the exploration boom peaked in 1988. During this period of time however the emergence of a significant presence of non-formal mining activities by local people in areas taken up by major exploration companies became prominent. At Marsuparia in Central Kalimantan for example, the contract of work area was deluged by village people from all over the province who mined the main reef using a simple but very effective method. Fires were built at the base of the main horizon and then water was thrown over the heated rock causing it to crack. Locals carried the fragmented rock to well-constructed "stampers" where the ore was crushed and then passed over cloth towels into baskets. Ultimately the fine crushed rock was panned for the gold.

Simatupang estimated that in the mid-eighties there were in excess of 10000 people in Kalimantan and Sumatra earning a modest living using traditional methods to extract gold. These methods include traditional, manual, partly unregistered "dulang" or "pan" washing of surface elluvial, alluvial and weathered primary deposits. Traditional dulang operations extend to about 2m below the surface and throughput is around 0.5–1.5 $m^3$ per day.

Landmark recently visited non-formal gold mining operations in West Kalimantan where the miners were claiming recovering 1–3 grams per day. At one "hard rock" mine he estimated 7000 tonnes per month of material was being mined by very simple methods.

There have been no medium-large alluvial gold mine that has existed for more than a few years since Independence.

Today the matter is complicated by the use in the non-formal mining sector of high powered gravel pumps and pressure hoses to mine alluvial deposits with significant impact on the environment. Where mercury is used to extract gold from ore, spillage of that metal has become an environmental hazard.

## Diamond

Diamonds in Indonesia have been known since antiquity but only from the island of Kalimantan (which was formerly known as Borneo). In particular diamonds are found in the provinces of West Kalimantan (Landak, Kapuas and Sekayan River areas), Central Kalimantan (Upper Barito, Sampit, Kapuas and Melawi Rivers) and South Kalimantan (Martapura, Cempaka and areas to the north and east).

One derivation for the name "Kalimantan" suggests a contraction from the Malay words "KALi eMAS daN inTAN" ie "Rivers of Gold and Diamonds"!

No undisputed primary source diamonds have been found and the source of the alluvial deposits is unknown. It has been suggested that perhaps the pipes were located in India prior to the Gondwanaland breakup. The deposits may be alluvial fans, surface or sub-surface palaeo river channels, or currently active present-day river gravel deposits. These alluvial deposits, which may represent the product of several cycles of erosion, have been worked by the local people using unsophisticated methods for centuries. The best-known areas are those at Cempaka and Martapura near Banjarmasin, the provincial capital of South Kalimantan. The Government has established a diamond cutting centre at Martapura to encourage local small scale industry. Current offficial production figures are low by world standards, only 15000–20000 carats per year, but total production from all sources may be several times these figures. The proportion of gem, near gem and industrial grade diamonds in the deposits are not known. The largest diamond found to date is the "Trisakti" of 166.8 carats, later cut to a stone of 50.35 carats.The earliest records indicate several diamond trade routes out of Kalimantan to Canton (China) and present-day Hanoi (Vietnam) from the 3rd–1st century BC. Porcelain fragments in some alluvial workings have been dated as Sung Dynasty (960–1279 AD).

There are no formal diamond mines of any size in Kalimantan even though there have been numerous attempts over the past one hundred years to find a primary source or an alluvial deposit of economic magnitude. Currently, Kalimantan is receiving another boost in diamond exploration particularly for offshore deposits.

Diamond production today is still in the Pertambangan Rakyat or non-formal sector. Villagers may pan for diamonds in the water course or groups may sink small shafts several metres deep into alluvials.

At one village the writer visited in Central Kalimantan in 1983 (Sungai Gula-Sugar Creek) the level of prosperity was very apparent, due to the main activity of diamond mining and trading.

## COAL

In 1995 the major production of coal in Indonesia was from Government and private non-government contractor mines. Production from those sources ranged from 665000 to 10200 000 tpa.

Nine national private companies outside of the contractor system produced 4220 000 tpa in the range 5520 to 800000 tpa.

Ten registered KUD's produced 229000 tpa in the range 350 to 81900 tpa. The non-formal/illegal sector may be producing as much as 1.5 million tpa

### Sand, Gravel and Aggregate

Due to the high level of construction in major cities like Jakarta there is strong demand for sand, gravel and construction aggregate. Production figures are difficult to quantify but daily rates of 50 to 750 tpd can be expected. At the lower end of the scale non-formal mining is active with attendant environmental and health hazards.

### Tin, Kaolin, Industrial Minerals

Production from Government companies and national private (KP), KUD, Pertambangan Rakyat activities.

## ENVIRONMENTAL RESPONSIBILITY

In Indonesia awareness of the relationship between the mining industry and the environment can be divided into three periods.

### The Colonial Period

Mining activity was on a relatively smale scale and the Government was concerned for the safety and health of the colonial miners to a certain extent. Impact on the environment was relatively limited and minor.

### Pre-Environmental Awareness in Mining Development (Since 1945)

Mining activities increased markedly and were mostly focussed on supporting the economic growth of the nation. Thus mining activities often resulted in considerable detrimental effects to the environment. As well there were often conflicts between mining development activities in other sectors e.g. Forestry, Agriculture, Transmigration (Human Resettlement Programs), Nature conservation etc.

## Mining and Environmental Awareness

Promulgation of the Act "Basic Provision for Management Environment" Law No 4 of 1982 stimulated the mining industry to a greater awareness of the impact of its activities on the environment.

Article 16 of that Act states:

> "Each development which causes a major impact on the environment should make an IEA (ANDAL) i.e. Environmental Impact Assessment"

This Act was later implemented by Government Regulations pp 29 of 1986 and pp 51 of 1993.

In 1990 the Government instituted BAPEDAL (Badan Pengendalan Dampak Lingkungan), an agency which manages the assessment of environmental impacts and monitors progress.

A mining project in Indonesia must undertake EIS (Analisis Dampak Lingkungan/ANDAL) to be completed with an RKL (environmental management plan) and an RPL (environmental monitoring plan).

At the November 1993 "Environmental Aspects of Mining in Indonesia" conference held at Jakarta, a number of issues relating to mining and the environment were identified. It was generally agreed that Government, National Private and foreign operated large scale mining activities (which are supervised by the Department of Mines and Energy) were not a great concern with regard to rehabilitation of the environment and its funding. However, the small scale mining activities were another matter and these were under the supervision of the Provincial authorities not the Department of Mines and Energy.

The technically unplanned operations of KUD/Village cooperatives, Pertamabangan Rakyat/Peoples mining, unregulated family alluvial mining, illegal mining under the guise of Pertambangan Rakyat were a particular problem.

This was because the people and organisations concerned generally did not have the inclination, "know-how" or funds to properly plan their activities to minimise damage to the environment in the first place or to ultimately repair the damage to the environment caused by their activities.

It was suggested that a levy could be made on the miners/ organisations concerned based on the tonnages extracted to provide a form of Security Deposit to provide funds for rehabilitation purposes. But in a response to that suggestion it was noted that in the past in Indonesia it was a historical fact that when such funds had been collected by the Provincial authorities for certain purposes they had been used for more urgent requirements.

## ISSUES FOR THE SMALL/MEDIUM SCALE MINING SECTOR

- Preservation of the environment and rehabilitation after mining completed.
- Provision of suitable technical training for local people involved in mining.
- Health and safety of persons involved.
- Feasibility studies and appropriate financing
- Small scale miners working in harmony with large scale activities. Appropriate regulation.
- Social issues. Working in harmony with other sectors e.g. agriculture, forestry, fisheries.

### REFERENCES

Anon 1988. Cooperatives obtain licences to operate coal mines, Business News (Indonesia), No 4711 p. 6.

Bambang Yunianto et al. 1995. The implementation of small scale mining (PSK) cooperation through miners participation, Indonesian Mining Journal, Vol 1, No 3, October, pp 74–77

Landmark, J. 1996. "People's Mining in Indonesia" Unpublished report as part of the fullfillment of the program of a Vincent Fairfax Ethics in Leadership Award.

T A Nurwinakun & Satry Nugraha, 1993. Legal Aspects for Foreign Investor in Mining Development in Indonesia.

Osman, A.H. (Tony) 1989. Diamonds in Indonesia Resource Review, September, pp 27–28.

Rachman, Ir. 1988. The Role of Cooperative in the mining industry in Indonesia, Mining Conference 1988. "Mining Prospects and Challenges in Indonesia during the Fifth Development Plan (Repilita V) Jakarta, September 28–30, 1988.

Robertson, G. 1988. A case study of a medium sized mining operation, Mining Conference 1988 Mining Prospects and Challenges in Indonesia During the Fifth Development Plan (Repilita V) Jakarta, September 28–30, 1988.

Simatupang, M. 1986. Gold Alluvial Challenges in Indonesia Conference, Indonesian Mining Industry General Review, May 7–9, 1986, Jakarta

Simatupang, M. 1988. Recent Development in Illegal Mining, Mining Conference 1988, Mining Prospects and Challenges in Indonesia During the Fifth Development Plan (Repilita V) Jakarta, September 28–30, 1988.

Simatupang, M. and Wahju, B.N. (Eds.) 1994. Environmental Aspects of Mining in Indonesia, The Indonesian Mining Association, Proceedings of the IMA/AusIMM Two-Day Seminar Environment. Jakarta, November 9–10, 1993.

van Bemmelen, R.W. 1949. The Geology of Indonesia, Government Printing Office, The Hague, The Netherlands.

Watters, R. 1995. Artisanal and Illicit Mining-Saran Gold Area, West Kalimantan, Indonesia unpubl report, Watters & Associates Pty Ltd.

# MARBLE MINING IN SMALL SCALE SECTORS IN INDIA—PROBLEMS, POSPECTS AND SUGGESTIONS

*A. Bhatnagar\* and S.K. Mukhopadhyay\*\**

\* Assistant Porfessor, Dept. of Mining Engg. College of Technology & Agricultural Engineering, Udaipur 313001, Rajasthan, India
\*\* Associate Professor, Department of Mining Engineering, IIT, Kharagpur 721302, India

## INTRODUCTION

Mineral resources are the back-bone of the country's economic development. Minerals are prime catalysts to the industrial growth and economic regeneration. Among various minerals, marble has come to stay today as highly preferred dimensional stone for various architectural uses due to its easy workability, durability, high resistance to weathering and aesthetic qualities like attractive colours. textures and designs etc. The mining of marble is still a small scale sector where mining is performed by private mine operators. The states of Rajasthan and Gujarat are main producers of marble in country with approximately 95% of total production. The marble mining in India proposes a good future but at the same time it is plagued with numerous problems generally because of its young age. The mining is yet not organised and measures, are essential to be taken, for planned, safer, scientific, economic and conservation friendly extraction of marble.

## PROSPECTS OF MARBLE MINING IN INDIA

The marble mining in small scale sector possesses great prospects mainly due to large reserves of good quality marble and export as well as domestic market potential. The need of the hour is to develop this sector by taking collective measures. The prospects of marble mining sector in India can be estimated on following aspects:

### Reserves

India possesses approximately 1200 MT of good quality reserves of various shades and designs patterns which suggest the enormous potential and offers challenging task to entrepreneur. The only need is the development of scientific and planned operations.

### Availability of Manpower

Our country prossesses a distinct advantage over competitive countries in the international market in terms of availability of large manpower at relatively lower cost, which may be helpful to increase the overall growth of the sector.

### Export Potential

The demand of marble in international market is increasing every year. India has the reserves and quality of marble to fulfill that demand.

### Infrastructure Facilities

Althrough India does not have the level of infrastructure facilities which are required but we are fast developing in this aspect too. India possesses following good aspects with reference to infrastructure facilities:

a) Reasonably good transport facilities.
b) Good banking facilities.
c) Skilled manpower.
d) Political stability.
e) Stable system.
f) Government's favorable intentions towards export.
g) Uninterrupted mining is possible throughout the year.

### Locational Advantages

India is strategically situated to cater the markets like Japan, Far East and Middle East. The only problem with India is its distance from Europen countries where Italy has slight edge over India and also the Indian products are costlier than Italian products because of losses in mining and processing. This aspect can be improved by scientific mining with reduced cost, higher percentage of extraction and by encouraging large volume transactions.

## Government Policy

Government is determined to develop this sector and taking suitable steps for this. Large number of entrepreneur are attracted towards this sector because of its potential and attractive policies of government. This will encourage the operators for organised and large volume mining.

## PROBLEMS AND SUGGESTIONS

Although the present status of marble mining in small scale sector is not disappointing, but to increase the overall growth and to fulfill the export requirements, we have to find out the problems faced by this sector and their countermeasures. Here we tried to formulate some common problems faced by marble mining sector in India with special reference to Rajasthan and Gujarat. The problems and suggestions are discussed in following classes for simplification:

Technical Problems.
Financial Problems.
Marketing problems.
Problems in Governmental Procedures and Formalities

## Technical Problems

*Problems*

Prospecting and location areas

Searching and location of the potential area is very tedious job because of infancy of present prospecting procedures. Although most of the part of country is covered by GSI and state bodies but details are not readily available since marble is minor mineral, low priority is given by exploration agencies. The procedures used, presently, for locating the potential area leads to develop pseudo-scarcity of areas causing increase in illegal practices like premiums, pressure tactics etc.

Exploration and evaluation of deposits and quality of stone

The usual procedure of evaluation of deposits, after identification is also primitive and inaccurate, which evaluate the deposits by assuming dimensions by past experiences and the quality is evaluated by polishing any sample. Due to lack of knowledge about deposit prior to mining causes high risk to the investment. Therefore finance is available for few purposes.

Problems in quarrying

a) Shortage of skilled personnel.
b) Underdeveloped technology.
c) Lack of training facilities.
d) Small size of quarries.
e) Lack of transport and communication facilities.
f) Uncertainty in government policies in licensing of quarries restricts long term planning and investments.
g) No prior information about the deposit, presence of cracks makes everything dependant on experience of persons in quarry.
h) All techniques are developed abroad. Although they are widely used but still they are not developed according to Indian conditions causing loss of energy and resources.

Problems in machinery and tools

a) The equipment, tools and machinery for mechanised quarrying are not manufactured in India and are imported. This aspect ultimately discourages mechanised and scientific quarrying.
b) Similarly, the loading machinery like cranes etc, are not manufactured indigenously, hence local industries are using general purpose equipment available.
c) Low explosives for blasting in marble industry are not readily available in India, which limits the use of blasting for quarrying.
d) Although chain saw and wire saw are readily available now but monopoly and import procedures cause disinterest to mine operators to adopt mechanised quarrying.
e) High cost of energy and tools, makes the mechanisation expensive.
f) Sometimes water is not available in plenty due to dry weather in the most of the areas.
g) Lack of adequate knowledge, about modern machinery, to mine operators which forces them to use conventional methods and equipment.

Problems in marble processing

a) Equipment available are not of best quality hence reduced productivity.
b) Lack of availability of spares of important equipment.
c) Lack of coordination between quarrying and processing industries.
d) Lack of quality control reduces export.
e) Lots of waste production in dressing of marble blocks. This cause pollution as well as loss of valuable marble.

Conservation and environmental problems

a) Percentage of extraction is very low in most of the mines due to lack of mechanisation, causing wastage of valuable of resources.
b) Some mines are capable of producing small marble blocks due to deployment of small machinery. This small block mining causes loss revenue which may be additionally obtined by production of large sized block.
c) Siliceous contents of stone causes real threat of dust pollution in surrounding areas.
d) Dumps of waste slurry from processing plants, which are near townships cause lot of dust pollution.
e) Large scale deforestation in potential marble areas will certainly have bad effect on ecology.

*Suggestions*

1. Intensive survey program should be developed by government for locating potential areas. The information concerning requirements of industry, value of mineral and potential of applications should be given attention in program.
2. For evaluation of deposit, a detail procedure should be followed. The first step to collect the information of following parameters.

a) Joint pattern
b) Marble boundary and interaction with host rock
c) Gradient
d) Age of formation—younger formations are less disturbed
e) Weathering
f) Overburden
g) Access from main road
h) Spilitting pattern
i) Colour of marble
j) Expectancy of large size block during mining
k) Compressive strength
l) Water absorption
m) Bulk density
n) Polish taking and retaining capacity
o) Presence of impurities
p) Sawability, workability, machinability, abrasiveness and spallability of marble.

Based on these, the deposit and quality of stone can be evaluated and following inferences can be drawn:

a) Total quarriable stone
   b) Recovery
   c) Size of proposed marble blocks
   d) Productivity
   e) Cost of quarrying
   f) Extent of mechanisation
   g) Uniformity of marble quality.

3. The dimensional stone mining should be taken up as separate specialisation in technical studies.
4. Proper training centers, like any other mining, should be established.
5. Research for planned and scientific mining should be encouraged. Fields should be optimised for raising the performance, reducing energy consumption etc.
6. Proper interaction between research and industry should be developed.
7. The period of lease allotment should be increased to facilitate long term planning.
8. The lease area should be increased to accommodate mechanisation and investment.
9. Manufacturing of equipment, tools, machinery and spares in India should be encouraged.
10. Standard specifications should be decided for all equipment used in mining and processing industry to increase uniformity in operations.
11. Proper power supply should be planned in remote areas.
12. Centres, local chapters can be encouraged to increase dialogue and interaction between mine operators as well as technocrats.
13. A proper quality control in processing plants should be developed, as in other countries, to exploit the export demand.
14. Proper maintenance program should be laid.
15. Countermeasures for all environmental impacts due to mining should be undertaken, like backfilling, revegetation, proper dumping of waste, secondary use of waste and slurry and use of boulders and small blocks.
16. Mine operators should be encouraged or forced by government to employ trained and technical persons in administration and as well as in operations.
17. Seminars, symposiums, workshops should be held regularly to acquaint the mine operators with new trends in method of extraction and developments in machinery in country as well as abroad.

## Financial Problems

*Problems*

1. The marble mining is very cost effective if large scale mechnisation is to be done, but the source for finance is limited and due to high risk involved the financial institutions are not liberal and are very selective.
2. Mining industry is not recognised as. industry by government; lots of facilities provided by various agencies, are out of bounds to this industry.
3. Much of the equipment used in mining is imported. The lack of financial support cause disinterest in mechanisation.
4. Lack of incentives to competent mine operators.

*Suggestions*

1. Government should take proper steps for financial help for mining of marble.
2. Increased participation of financial institutions may increase overall performance of marble industry.
3. Import facilities to be increased for essential equipment and if may be good for overall development of industry. The export generated by this will accommodate the loss of revenue in import facilities. But at the same time, indigenous manufacturing of those equipment should be encouraged.
4. Government may give tax holiday for some duration to export oriented mines and incentives in any form, to mechanised mines.

## Marketing Problems

*Problems*

1. Lack of standards for size, thickness and quality causing haphazard marketing practices.
2. Marketing is still underdeveloped and poses numerous stages of selling upto finished product. This aspect of too many transaction from ROM to finished product causes lots of malpractice, less attention to quality control, increased waste production and lots of compromises on part of small traders.
3. Due to lack of quality standards, same marble may have different selling prices. This aspect increases the role of brokers or middlemen:
4. The marble mining is typically seller's market, so mine operators have their whims, norms and preferences of selling i.e. to whom to sell and at what cost.

5. At the same time, preparation of standards for quality control is a tough job since the quality of rock vary from mine to mine, block to block and also within the block itself.
6. Due to lack of collectivity on part of mine operators and lack of quality control, it is an irony that sometimes industry is not able to fulfill bulk demand of same quality of marble.
7. Operators are satisfied with the present selling as they are getting huge profits, so there is no attention towards product upgradation, advertisement, innovations and healthy competition.
8. Sometimes due to petty interests, preference of mine operators delays the supplies, affecting the reliability of industry.
9. Export potential is not fully utilised due to unplanned marketing and lack of coordination between traders.
10. The mine operators do not follow the marketing norms by offering to sell on credit for some period like their overseas counterparts do, that reduces large volume business. But at the same time very few can afford to do so, as they require the money in rotation to be pumped back into mining because of almost nil financial assistance from any source.
11. Indian products, though superior in quality, are unable to secure a export market due to high production cost.

*Suggestions*

1. Although the quality varies from block to block but standards of finish can be established. Proper quality control and sorting according to quality of tiles at processing plant should be encouraged.
2. Large scale extraction, processing and marketing should be encouraged.
3. Proper coordination between mining, processing and marketing segments of industry will better the quality control and reduce the wastage.
4. Government's proper intervention, regarding provision of marketing platform, steps for creating better environment and measures to increase interaction between various segments of industry, may have good impact over marketing practices.
5. A combined directory can be prepared which can include name and address of mines, processing industries, marketers, technocrats and any other related to industry.
6. Development of an organisation or central body, may be helpful in better marketing practices. The body can endorse the deal, either cash or credit, to avoid the risk of payment and delivery. Also large volume transctions can be increased by this.
7. Trade promotion cells can be developed by the government.

## Problems in Government Procedures and Formalities

*Problems*

1. The licensing procedure is time consuming.
2. The quarry lease for short duration and of small size, specially in Rajasthan.
3. The government of India and various state governments have been unable to provide a sheltering roof to this industry. A lot of delay in export clearances, timely shipment etc. occur.
4. There is no systematic information system in mining sector. The producers and government agencies have no systematic records and compilation of data for guidance of new entrepreneur.
5. Procedure of demarcation is faulty. The lack of modern equipment and use of conventional chain and compass in survey causes lots of boundary disputes.
6. Large deposit of good quality marble falls under so-called green belt or reserve forest, specially in Rajasthan and Gujarat, where all those land are lying barren without any vegetation since long.
7. Mine operators are keeping the lease for years without mining in hope of large premiums. It is causing loss of revenue, royalty and overall development of industry.
8. While allotment of lease, the fact is not taken into account that the aspiring lessee will be able to mine it with good technique at higher percentage of extraction.

*Suggestions*

1. The procedures for various governmental formalities should be made easier.
2. Single window assistance of various procedural formalities may also be developed on the part of government, as being practiced in other industries.
3. The restrictions over size of lease should be removed, as there is no limit on size of lease in Gujarat. It will encourage mechanisation and long term planning. The duration of lease should also be substantial.
4. Government should come ahead and take necessary measures to protect and encourage mine operators for procedural formalities, technical feedback and training.
5. Since mine sites are generally in hilly region, modern equipment for survey such as theodolite etc. should be used.
6. Central government should be persuaded for exempting potential areas from forest. For that purpose some environmental measures can

be made mandatory for mine operators to counter the loss to environment.
7. Inter-departmental issues involving departments such as forest, mining, revenue etc. should be sorted out expedititously.
8. The different varieties of marble available in India should be documented and scientific classification should be encouraged.

## CONCLUSIONS

The future of marble mining in small scale sector is very bright although there are numerous problems. The most important aspect of this sector is that, not a single problem is so grave that it cannot be solved by keen interest and dedication. India possesses large reserves with good domestic and international demand. In addition to these demand and reserves, by devleopment of new scientific extraction and processing techniques, this industry is bound to reach greater heights in the near future and develop as a potential and vibrant sector in our economy.

### REFERENCES

Anon.1993. "A Technological evolution and norm study in marble and granite industry - A case study prepared by Government of India for technological self reliance". July 1993.

Banthia, H.R. 1990. "Needs of R & D and innovative techniques for proper development of dimensional and decorative stone mining and processing"; Seminar on Dimensional and decorative stones of Rajasthan, Udaipur, April 26-27.

Bhatnagar, A. 1995. "Technoeconomic aspects of marble mining in Rajasthan"; Post graduate seminar, Department of Mining Engineering, IIT Kharagpur, Feb. 28.

Chowdary, T.V. 1992. "Problems and prospects of dimensional industry — An overview", Indian stone, pp. 5-21.

Mathur, A.S. 1990. "Some suggestions for development of dimensional and decorative stone industry In Rajasthan"; Seminar on Dimensional and decorative stones of Rajasthan, Udaipur, April 26-27.

Mohnot, V., Mohnot, M. and Jain, R. 1994. "Marketing of marble products: Problems, & prospects"; International Seminor on Marble industry, its problems & prospects; Udaipur, Sept., 25-26.

Sarangi, S.K. 1993. "Status of dimensional and decorative stones of Orissa - An overview"; Indian Mining journal, 32(4), pp. 33-38.

Vaish, A. and Parihar, A.K. 1994. "Marble and granite industry of Rajasthan: Its potential problems and prospects"; International Seminar on Marble industry, its problems & prospects; Udaipur, Sept. 25-26.

# RECENT TRENDS OF DIMENSIONAL STONE MINING IN INDIA

K.U.M. Rao[1] and A. Bhatnagar[2]
[1] Assistant Professor, Dept. Of Mining Engineering, I.I.T. Kharagpur 721 302, India
[2] Assistant Professor, Department of Mining Engineering, College of Tech. & Agri. Engg., Udaipur 313 001, India

## INTRODUCTION

The use of stone as a building and monument material is an ancient art and technology in India. Any stone specially cut, shaped to various sizes for use as a building material is called as "Dimensional Stone". Amongst various dimensional stones granite and marble have come to stay today as the highly preferred rock in terms of architectural design. It is mainly because of their inherent physical characteristics as well as their elegant appearance. Statistics reveal that the demand for dimensional stones is ever increasing. In view of this rise in demand, there is a considerable interest in improving the methods of extraction for higher productivity and better quality. However, method of mining imposes certain constraints, in the sense that, unlike the other commercial minerals, the dimensional stone mining is constrained to obtain the blocks of required dimensions, which are free from induced cracks. A variety of techniques are used to extract blocks of stone from quarries, which include water jet and flame jet cutting, drilling and blasting, and wire sawing with other conventional methods. However, the present trend in dimensional stone mining in India is to adopt diamond tipped cutting tools for the extraction of blocks. This method, in particular, has gained popularity in softer formations such as marble.

In dimensional stone mining much of the cost is involved in processing of mined out blocks. It consists of cutting of slabs from bigger blocks and tiles from smaller blocks. Diamond tipped cutting tools such as gang saws, disc cutters are the major consumables in this operation, which infact are very expensive. Therefore, any improvement in their total service life would result in increased productivity. In this paper, the attempt to increase the life of the diamond tipped tools by using a non-ionic polymer as an agent in flushing water is also discussed.

## COMMON QUARRYING TECHNIQUES

The dimensional stone mining involves dislodgement of large blocks from the rock mass. By and large overburden associated with granites and marble in India, is very shallow. Therefore, simple methods of overburden removal are adopted. Based on the degree of mechanization adopted in overburden removal, they are classified into: (a) manual, (b) semi-mechanised and (c) mechanised methods.

The manual and semi-mechanised operations adopt conventional methods of drilling holes with pneumatic jack hammer, followed by blasting operations, and for transportation cranes and dumpers are deployed. The mechanised operations additionally involve hydraulic shovels for overburden handling. After the overburden is removed the whole acitivity of mining is reduced to the extraction of intact blocks with least possible damage to the surrounding blocks. This operation is in general known as delineation and there are different methods of delineation in vogue.

## METHODS OF DELINEATION

The operation of delineation is a skilled work which involves extraction of intact rock blocks with least disturbance to the surrounding rock mass. The different techniques adopted to delineate the block are: (a) Feathering, (b) Drilling and blasting, (c) Overlapping holes, (d) Diamond wire sawing, (e) Helicoidal wire sawing, (f) Flame cutting, and (g) Chain sawing (Ajay Kumar et al., 1995).

### A. Feathering

In this method, holes are drilled along the joints and other sides. The holes are drilled in the required plane at a very close interval, of say 15 cm in between them, while their depth depends upon the required size of the block and the drill availability. A wedge, with two metallic thin plates, called feathers, is inserted into the holes. The wedges are hammered uniformly one after another with a sledge hammer until a crack is formed along an artificial plane of weakness.

### B. Drilling and Blasting

This method involves drilling of closely spaced holes, at 10–15 cm spacing, on the proposed face and then blasting them by low strength explosive, initiated by detonating cords, while the holes are either water stemmed or air stemmed and plugged at the mouth.

## C. Overlapping

The blocks are dislodged by cutting a slot by drilling overlapping holes.

## D. Diamond Wire Sawing

The proposed slot is cut under the influence of vertical and normal forces at the interface, transferred by the cutting beads which forms a closed loop round the rock blocks. The beads contain on their surfaces, the impregnated diamond grits, which produce dragging action during motion. The loop of wire is formed by drilling suitable holes on the proposed slot boundary. The cutting speed and the tension to the wire are given by a drive-pulley arrangement.

## E. Helicoidal Wire Sawing

An endless helicoidal wire with single, double or triple strand, is made to run over sheaves under sufficient tension. The aqueous siliceous solution fed along the wire, acts as abrasive material, cuts the slot along its path. This plant consist of an engine, a helicoidal wire, a tension device and two idle pulleys which will press the moving wire against rock to be cut.

## F. Flame Cutting

The flame cutting is preferred when blasting techniques become ineffective for higher production. The flame cutting equipment consists of burner, fuel mixture tube and separate inlet for air, kerosene and oxygen. The fuel mixture is ignited at the burner nozzle and a narrow sharp flame is produced. The flame is held over the rock which causes a slot in rock due to intense heating.

## G. Chain Saw

The chain saw is similar to coal cutting machine, and it has a special endless chain mounted on a jib. This is a mobile unit by which slot can be cut by placing it at desired place. Generally it is used in conjunction with diamond wire saw for making initial slot.

## RELATIVE ASSESSMENT OF VARIOUS DELINEATION TECHNIQUES

The primary objective of all the various delineation techniques is to increase productivity, with due consideration to the conservation of fast depleting dimensional stone reserves. Many efforts are being made towards this objective and consequently today we have fast developing advanced techniques like, flame jet cutting, water jet cutting etc.

Of the existing methods feather and wedge method is followed by large number of mines owing to its simplicity and mobility. Nevertheless, it suffers great losses which leads to lower percentages of extraction. Further, it can handle only smaller sized blocks. This handicap is primarily eliminated in the drilling and blasting method, which also improves production. In blasting the damage to the rock is due to the development of cracks. Therefore, great care is required in operations under this technique. Attempts are being made to reduce the crack propagation into the intact rock mass. Both the diagonal slotting and Finnish methods, known as smooth blasting techniques, are the improvements in blasting operations in dimensional stone mining.

The promising methods of delineation are the helicoidal wire sawing as well as the improvement over this, the diamond wire sawing. Primarily, both these approaches gives better finished block with smooth surface and the block remains intact. However, the helicoidal wire sawing involves higher initial capital investment and also the lower cutting rates (1–1.15 sq. m per hour) adds to production costs; therefore, the diamond wire sawing is slowly replacing the helicoidal wire sawing. Diamond wire sawing is very effective in marble and sandstones.

The widely adopted techniques of delineation in India are feather and wedging techniques and diamond wire sawing. Figure 1 gives a comparison of all the delineation techniques.

The blocks mined out require processing before shipment. The operations in the processing plant, comprises, cutting of uneven sized blocks to size by diamond disc cutters, cutting of slabs and tiles from various sized blocks by different types of gang saws. Principally, diamond tipped cutting tools occupy a major proportion of equipment in the processing plant.

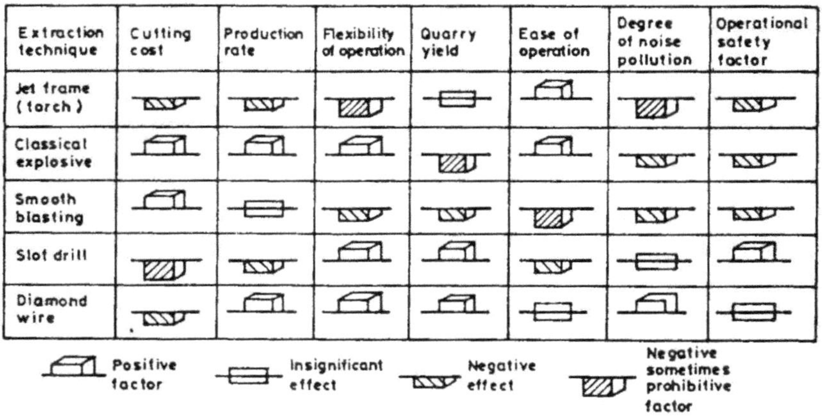

Figure 1. Comparison of main erxtraction techniques

239

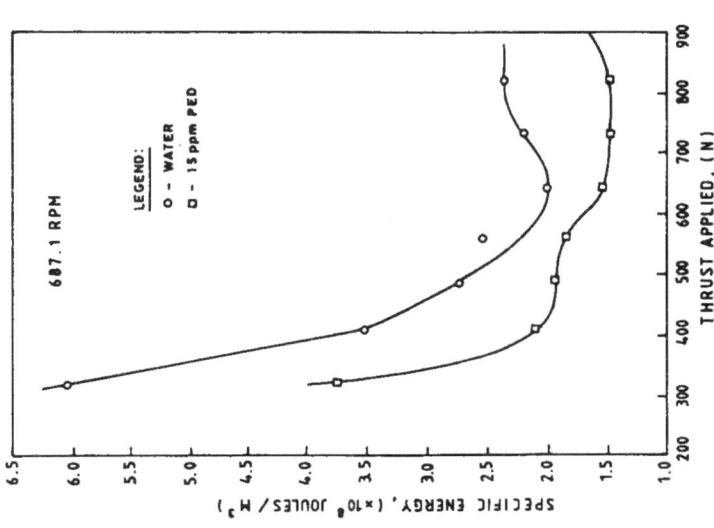

Figure 2(b). Influence of thrust on specific energy consumption for different flushing media. (Rock type: Marble).

Figure 2(a). Influence of thrust on specific energy consumption for different flushing media. (Rock type: Sandstone).

To improve the life of the diamond tipped cutting tools, extensive laboratory scale experiments were conducted. Since the principles of rock fragmentation in all these types of mechanical cutting tools is same, the cutting by disc cutters is simulated to a diamond tipped impregnated core drilling operation.

## LABORATORY EXPERIMENTATION

All the laboratory experiments were carried out on a heavy duty vertical drill machine of 5 H.P. manufactured by Hille-werke Dresden, Germany. The assumptions made in these studies are that thin-walled impregnated diamond core drilling can best be simulated to the diamond wire sawing, and diamond disc cutting. The sawability is therefore a synonym to the drillability of rocks by diamond drilling (Rao et al., 1994).

## ROLE OF PEO ON THE PERFORMANCE OF DIAMOND DRILLING

One of the main objectives of the present investigations was to establish and understand the necessary conditions of chemical control of drilling performance.

From all the drilling trials with various combinations of applied thrust and rotational speed, the effectiveness of PEO as a flushing medium is that: (i) the magnitude of the torque generated at the bit-rock interface, is less in all the rock types; (ii) Similarly, the penetration rates achieved are higher when PEO is used in flushing water.

In the present investigation, an increase of around 10% to a maximum of 50% improvement in the penetration rate was obtained against 400–600% improvement reported by USBM. Therefore the concept proposed by USBM regarding the influence of PEO on the penetration rates cannot be contradicted. It is therefore inferred that with the addition of PEO, in a given rock, the surface charge gets neutralised, which is one of the reasons for enhanced penetration rates. Secondly, PEO solution also behaves as non-newtonic fluid. This reduces the thickness of the fluid layer at the rock surface, which increases the efficiency of energy transfer by a drill resulting in a greater cutting efficiency.

According to another school of thought (Engelmann et al., 1987), drag reduction in the polymer solutions is the phenomenon whereby extremely dilute solutions of high molecular weight exhibit lower frictional resistance to flow than the pure solvent.

The observed lower torque values in the present investigation supports this phenomenon of reduction in the turbulent friction.

## REFERENCES

1. Ajay Kumar, L., Srinivas, K. 1995. Methods of Mining commercial granites — a critical review and suggestions", National Symp. on Commercial granite, Madras, Feb. 4–5, pp. 197–199.
2. Rao, K.U.M and Misra, B. 1994. "Design of a Spoked Wheel dynamometer for simultaneous monitoring of thrust and torque developed at bit-rock interface during drilling" Int. J. Surface Mining Reclamation & Environment, Vol. (8) pp. 146–147.
3. Engelmann, W.H., Watson, P.J.; Pahlman, J.E., "Zeta. 1987. Potential Control for Simultaneous Enhancement of Penetration Rate and Bit-life in Rock drilling", U.S.B.M. RI 9103; pp. 18.

# ARTISAN MINING OF COAL IN THE GARO HILLS, MEGHALAYA

*D.K. Jain* and A.K. Sural***
*Chief General Manager, Central Mine Planning & Design Institute, Ranchi, India
**Deputy Chief Mining Engineer, Central Mine Planning & Design Institute, Ranchi, India

## INTRODUCTION

All the three hill divisions of the state of Meghalaya are endowed with the deposits of coal, found in a large number of small coalfields scattered over the hills and valleys amidst forest cover. The major share of coal reserves are located in the Garo hills which lie on the western-most part of the state, bounded by planes of Bangladesh on the west and the south.

Coal mining in the Garo hills started in late seventies producing insignificant quantities of coal from a few mines. Over the years, production from these artisan mines grew at a fast rate with increasing number of mines here and there. Today, a large number of such artisan mines, each of which produce a few tonnes of coal each day, can be seen alongside hills located in far-flung areas, where basic infrastructural facilities are totally absent. Despite this fact these artisan mines cumulatively produce a large quantity of coal, which plays an important role in the economy of not only the division but the state as a whole.

The mining practices are very much primitive, which do not take cognizance of even basic safety requirements or the need to exploit coal in a conservative manner for the business to go on. This paper brings forth some information concerning coal deposits and their exploitation.

## GEOLOGICAL APPRAISAL

The Garo hills, forming the western part of the Shillong plateau is flanked by Tura range stretching WNW-ESE over a length of about 100 kms. The entire range, within which the coal deposits are located, forms a steep escarpment and rugged hilly topography with prominent peaks rising up to

1026 m. Average elevation of the area varies from 150 m to 300 m above mean sea level.

The area is characterised by deep ravines with steep hill slopes on either side. Perennial rivers viz. Simsang, Maheskhola, Mahadeo, Kanai etc. with their many tributaries flowing generally southernly, drain the area into the plains of Bangladesh.

The higher reaches of the hills in the entire coal-bearing area is covered with thick forest consisting of deciduous and semi-evergreen species and a large variety of scrubs. The foothills, in patches, are devoid of vegetation as a result of tree felling and the practice of shifting cultivation.

The nearest important railhead is *Jogigopa* about 120 kms from the Nongwalbibra—the principal centre of mining activities in the Garo hills. Nongwalbibra can be approached by a narrow metalled road from *Tura* (about 100 kms) or from Baghmara (about 70 kms), the district headquarters of West Garo hills and South Garo hills respectively. The region experiences heavy rainfalls during monsoon spread over 5 months a year.

The coal deposits of Garo hills relate to Eocene age and forms a part of the Tura sandstone formation. Minor deposits, belonging to Lower Gondwana sediments, also occur in the western part of the Garo hills in few lenses only, which do not bear any economic importance. The tertiary deposits are confined to the southern slopes of the plateau adjacent to the plains of Bangladesh.

In the Garo hills, six small coalfields have been established in the course of regional exploration work carried out by GSI. Exploration activities were mostly limited to geological mapping and scanty borehole drilling at few places, except in West Darangiri coalfield, where major part of the area has been covered by more closely spaced drilling.

These coalfields bounded by latitudes of 25°12' N and 25°44' N and longitudes of 89°58' E and 90°58' E, are spread over an area of approximately 470 sq. kms in patches, isolated by washouts. Coal deposits in these coalfields exhibit simple structures.

Of the six coalfields, mining activities are confined to only three viz. *West Darangiri, Siju* and *Karaibari*. Rest of the coalfields are lying virgin, either due to inaccessibility or because of their location within reserve forests and biosphere reserve.

Brief outline of these three coalfields are presented below:

**West Darangiri Coalfield**

This is the most important of all the coalfields of Meghalaya. Occupying an area of about 47 sq. kms., close to *William Nagar*—the district headquarters of East Garo hills. This coalfield is also the best established by close drilling and exploration. The area can be approached by a metalled road connecting William Nagar to Baghmara, which passes through Nongwalbibra — a small township, surrounding which mining activities are in progress.

There are four coal seams ranging in thickness from 0.1 m to 2.98 m, of which three seams are impersistent. The second from top, named *Darangiri seam* is persistent over most of the coalfield and varies in thickness from 0.30 to 2.98 metres. This seam has economic importance and is being mined.

The seams are somewhat flat with rolling dip ranging from 2 degrees to 5 degrees. The coalfield is traversed by one fault, throw of which varies from 15 to 40 metres.

Darangiri seam and other seams are exposed at several places on the steep hill slopes, all along the river channels of *Simsong, Rongju* and their feeders. The seams are closely spaced with partings varying in thickness from 2 to 20 metres. The coal is friable in nature and rather easy to cut by hand pick. Immediate roof of Darangiri seam is composed of alternating clay and sandstone. The floor is composed of clay, silty clay and carbonaceous shale.

The top weathered mantle is highly permeable, which allows a significant part of the rain-water to percolate into the mines, having little top cover.

Proximate analysis indicate that coal of Darangiri seam contains high moisture (4 to 18%), high volatile matter (34 to 58%) and low to moderate ash (1 to 21 %). Total sulphur content varies from 1 to 4%, a larger part of which is organic in nature. Calorific value of this coal ranges between 6470 and 8225 kcals/kg. Unlike coal from Jaintia hills or Khasi hills, this coal does not exhibit caking properties.

Coal reserves in this coalfield amount to 127 million tonnes, of which about 65 million tonnes fall under proved category.

This coalfield is under active exploitation and account for more than eighty per cent of coal produced in the Garo hills.

**Karaibari Coalfield**

Karaibari coalfield is located alongside the *Garobada-Singrimari* road and falls within the administrative block of *Shelsela* of West Garo hills district. The headquarters of the district — Tura is about 32 kms on the south-east.

Evidences of coal occurrences are available over a stretch of 15 kms. Coal exposures varying in thickness from 0.20 to 1.30 m along hill slopes have been encountered. Only marginal prospecting work has been carried out in this coalfield, which does not enable one to corelate the seams.

Coal extracted from mines in this area contain a large proportion of lumps indicative of being harder coal compared to Darangiri coal. Roof and floor rocks are also more compact compared to Darangiri coalfield.

The coal is dull, bown in colour and breaks with concoidal fracture. Analytical results obtained from few channel and run-off-mine samples indicate that the coal is high in moisture, low to high in ash and high in volatile matter content. Sulphur content is also quite high (up to 6%). Calorific value of coal varies from 6500 to 6800 kcals/kg.

No reserve in this coalfield has been estimated in absence of reliable exploration work. Mining in this coalfield is being carried out to a limited extent.

## Siju Coalfield

Siju Coalfield, covering a wide area of about 225 sq. kms. is located 25 kms south-east of Darangiri on the William Nagar- Baghmara road. The coalfield is also referred to as *Hidden Coalfield* as the coal-bearing Tura formation is overlain by a 100-metre thick limestone capping.

Physiographically, the area is characterised by rugged and undulating topography with hummocks at places and is marked by steep escarpment at the southern margin of Tura range. Much of the area depicts typical features of limestone cover including caverns and sinkholes. Major part of this coalfield is located within reserve forests and Balphakram biosphere reserve.

Only limited exploration has been carried out, through mapping except for an area of 7 sq. kms, where more detailed investigations have been made. Two seams have been identified at depths ranging between 93 metres and 258 metres. The upper seam is 0.5 to 3 m in thickness. The lower seam, separated by a parting of 7 m to 136 m over the property, is 0.6–1.5 metre in thickness.

Analysis of channel samples and spot samples show that the coal contains high moisture, low to high ash and high volatile matter. Sulphur content varies from 1.2 to 3.5 per cent. Calorific value of coal is as high as 7900 k.cals/kg to 8700 k.cals/kg.

The coalfield is mostly virgin due to its location within reserve forests and inaccessibility. Only the northern part of the coal deposits is being exploited at *Chokpot* area, which is 64 kms from Tura. This area can be approached from *Adugiri*, 23 kms from Tura on the *Dalu* road.

In absence of closer geological survey not many details of this potential coalfield is available. Reserves, estimated under inferred category is 127 million tonnes.

Other coalfields viz. *Rongrengiri. Balphakram-Pendengru and East-Darangiri* are lying virgin. Except for Rogrengiri which lies within reserve forest, the other two coalfields are almost inaccessible. Total coal reserves in Garo hills, so far estimated amounts to 359 million tonnes, out of total estimated reserves of 460 million tonnes in the state of Meghalaya.

Coalfield wise coal reserves and their quality characteristics as obtained from few samples analysed are summarised in Table 1.

Table 1

| Coalfield Name | Estimated Reserves Million tonnes | Moisture Percent | Ash Percent | Volatile Matter Percent | Calorific Value* k./cals/kg. |
|---|---|---|---|---|---|
| Karaibari | Not assessed | 6.1-29.0 | 3.1-29.0 | 35.3-54.0 | 6255-8635 |
| Siji | 125 | 1.7-11.3 | 1.5-33.4 | 32.6-56.7 | 7895-8720 |
| West Darangiri | 127 | 4.6-18.0 | 0.8-21.8 | 34.2-47.5 | 6470-8225 |
| East Darangiri | Not assessed | 4.5 | 6.7 | 52.3 | 8,550 |
| Rongrengiri | Not assessed | 8.3-17.4 | 2.8-17.4 | 40.2-50.6 | 5495-6240 |
| Balphakram-Pendengram | 107 | 1.6-11.1 | 3.0-14.6 | 33.1-53.1 | — |

*On pure coal basis.

## THE MINING SCENARIO

History of coal mining activities in the state of Meghalaya is not very old. Prior to nationalisation of the Indian Coal industry, coal mines, few in number began to operate in the Khasi hills for export of coal to Bangladesh. After nationalisation, demand on coal from Meghalaya took a big jump with expansion of export as well as domestic markets. Coal mining activities thus extended further—first into the Jaintia hills and then into the Garo hills. Coal mining activities, so long confined to the communities of few villages, began to grow at a fast rate with the involvement of fortune earners from other localities, entering into coal business.

In the Garo hills, mining is being carried out in three coalfields viz. Karaibri, Siju and West Darangiri. While limited mining activity is seen in Karaibari and north-western part of Sigu Coalfields, there has been mushroom growth of artisan mines in west Darangiri coalfield over the past few years.

### Customary Rights of Mining

Like in the Khasi and the Jaintia hills almost entire land property belong to the local tribal population, member of which, as provided under customary tribal law, also enjoy the right to mine and sale any mineral within the limits of their land property. The village chief, who is the trustee of all landed property in and around a village, may, against royalties payable, permit any member of the community to carry out mining operations within the area allotted to him. Owing to limited extent of exposed coal, where every body wants to carry out mining, the piece of land allotted to individuals is in the form of a narrow strip 10 m to 20 m in width across hill slope which run well into the hill. This factor results in closely spaced mine entrances with consequent problems.

## Lack of Infrastructure

Mining activities in West Darangiri coalfield surrounds Nongwalbibra, which is located far from the source of even bare minimum requisites for bringing about any cognizable change in the mining culture.

The only metal road passing through Nongwalbibra allows transport of coal to outwards destinations during dry months. Clusters of petite mines located within five kilometres are accessible only with difficulty. Some of the mines are connected to this metal road by foot tracks only, while some others, by make-shift pathways laid by local villagers, allowing small 4 tonne trucks to fetch coal from mouths of the mines to a stocking place by the side of the metal road.

Though a high tension electrical transmission line passes through the coalfield, the mining sites are not connected by electrical network. Civic amenities are almost non-existent. One has to depend on the township of Tura — 100 kms away, for raw materials required for any construction work.

## The Mining Community

Population density in the mining belts of the Garo hills is very thin. Small villages/hamlets are found located in a scattered manner amidst hills and forests. Majority of people earn their living through *Jhumming*, (shifting cultivation) and forest produce.

Local tribals show reluctance towards hard physical labour, such as, in mining. Almost entire workforce in the coal mines are immigrants from Bihar, Bengal and Assam. Skilled men as required for even the simplest form of mechanisation are not available around the area.

The workers are employed on piece-rate basis and a healthy worker usually earns good amount of money on daily basis. But due to stoppage of mining activities during monsoon (4–5 months in a year), large size of their families and high price of edibles, their living standard is very poor. Moreover, loss of job is common due to sickness and closer of mines for one reason or the other. Thus their living is generally unstable.

These workers live under pitiable conditions in temporary slums erected around the mines. Health and medical facilities are practically non-existent. Sufferings from malaria, jaundice and lung diseases are very common among the workers as well as locals.

## Environmental Status

Environmentally, the state of Meghalaya is considered as very rich but fragile. A large variety of flora all through the state can be found. In the coal belts of the Garo hills, the forests are dense, sometimes inaccessible and boasts some of the rarest varieties of animal life.

Very heavy rainfall, thin population, difficult access and absence of industries—all in combination has so long contributed towards distinguishing some of the areas with richness in terms of flora and fauna and pollution-free atmosphere. *Balphakram*, one of the finest biosphere reserves of the country, is located, close to the mining areas. In fact, a part of the mining fields fall within the buffer zone of the reserve.

The area exhibits large variations in surface contours, with major drainage channels cutting across the coal belt. The soil is generally loose and suffers from high degree of erosion when exposed. Deforestation of the densely covered steep hills in the lower reaches is being done for Jhumming as well as timbering and to some extent for facilitating mining operations. Mines under such uncovered areas are found to be most prone to collapse, followed by landslides, though in a small scale. The affected areas get eroded very fast due to torrential rain experienced during monsoon weakening the hill slopes and the process continues.

## MINING OPERATIONS

In Darangiri coalfield the main workable seam, i.e. Darangre seam varies in thickness generally from 1.2 metres to 2.5 metres. Significant quantities of coal is also mined in *Chokpot*, falling in the outlying areas of Siju coalfield. In these areas coal seams are found exposed in many places on hill slopes, making it possible to enter into the coal horizon without any earthwork. A large number of entries are made close to each other and belonging to different owners.

The seam disposition is almost flat. Immediate roof is composed of clay, fire clay and caronaceous shales, which does not hold itself when exposed. Attempts are made to drive the entries and galleries in an arch shape, with the crown of the arch, touching the immediate roof of the seam. However, slightest exposer of the roof, leads to leaching action by water present in the surroundings, and the roof flows in. During dry season, the immediate roof tends to part off from the strata above and in some cases falls abruptly.

The entries and the galleries are driven in random fashion, forming pillars of irregular shape and size. Maximum dimension of any odd-shaped pillar hardly exceeds 5 metres. Height of galleries equals the thickness of the seam and the width varies from 1 metre to 2.0 metres. Generally, supports are not used except in the portal area, where a few props are erected in some mines. Coal is won using hand pick which is the only mechanical appliance ever to be seen in a mine and the cut coal, loaded into cane baskets, is carried out of the mine manually and stacked close to the mine entrance. Each coal cutter is usually backed by two loaders, which form a team. A second set of coal loaders carries the coal to the roadside gathering points, again by cane baskets, carried on the back.

Quantity of coal produced by each team is measured in terms of wooden boxes filled by the team.

In some cases, where the mine sites are far from the motorable road and the terrain permit, the villagers construct make-shift roads connecting the mines, which allow light trucks (four tonnes capacity) to shuttle between the mines and the gathering point. The villagers collect levies from mine owners for using such connectors road.

The mines generally extend in bye to about 200 metres depending upon adequacy of natural ventilation. For lighting purposes, candle lights are used.

## Production and Productivity

Depending upon size of the mine, the lead distance and the possibility of finding a suitable buyer, the total employment in a mine varies from 10 to 40 per day producing 10 to 30 tes in a day. Each coal cutter is associated with 2 underground loaders. Each cutter cuts two to four tes in a day engaging himself form dawn to dusk. Underground man productivity of a mine, calculated at the works out to 1.0 to 1.3 tonnes

## Safety Aspects

The mines extend inbye as long as the natural ventilation permits. The limits generally vary from 100 to 300 metres. A cluster of mining holes, closely spaced amongst themselves are generally inter-connected with each other. Adequacy of ventilation solely depends upon direction and speed of natural air flow at the surface. Supports are hardly used except where there is visible danger of roof caving in or sometimes at the portal.

Presence of ignitable gas is unknown. Candle lights carried by miners are used for lighting purposes. Miners are unaware of personal protective wears.

In a bid to extract maximum quantity of coal at the earliest possibility, cross galleries are driven too close to the portal. Due to shallow cover over the cross galleries and very small size of pillars collapse of entrances are very common during monsoon. Once an entrance collapses, neighbouring entrances also gets affected. After the monsoon is over, the land mass above the affected areas, slides down, causing land degradation and subsequent loss of entrances.

As reported by many, accidents—mainly due to collapse of ground and often fatal, are common in the mines. Due to confined space underground, work posture is highly arduous and most of the miners suffer from serious back injuries. The workers of the mines complain that within five to ten years time their earning capacity reduces due to physical disablement arising out of hard labour and drudgery.

In these mines, adherence to the mining legislation or its enforcement appears to be totally absent.

## Conservation

As stated earlier, the mines comprise mainly, drivage of a few galleries inbye towards the centre of the hill, under which the deposit occurs. Due to lack of ventilation, long lead distance and at times, collapse of roof or the entrances, the mines are abandoned within short span of time. Thus numerous holes placed radially all along the periphery of hills are seen. With passage of time these entrances become unusable, and at the same time blocks the reserves lying in the central part under the hill. Though no estimates can be made of the rate of extraction, certainly, very large proportions of reserves are being lost for ever, as a result of such practices.

## SUMMARY

Although reliable records are not available, rough estimates show that the current production from some two hundred and odd miniature mines in the Garo hills, stands at about 1 million tonnes, and show an upward trend. The coal mining industry in this, industrially underdeveloped area, plays a major role in the local economy and employment of people.

The mining culture prevalent in this remotely located area may appear to be primitive. But, if one considers the background perspective of the area and its people, the system being practised is possibly relevant and has its own merits, under the existing socio-economic structure.

The points of concern are the low conservation rate and environmental degradation, resulting from unplanned and unscientific mining system and poor state of safety of the mines as well as the workers, which can be ameliorated significantly, with little technical assistance combined with efforts of the local socio-political organisations. The state of affairs demand urgent attention of the local people for protection of their own interests.

## ACKNOWLEDGEMENT

The authors are thankful to the management of CMPDI for the permission to present this paper. The views expressed in this paper do not necessarily reflect the views of the management of CMPDI.

# SECTION 5

# Economic Perspective

# COMPETITIVE EDGE OF MINERALS FROM INDIA's SMALL SCALE MINES IN GLOBAL MARKET

*C.S. Jha*
Mining Industry Consultant, Ranchi, India

## INTRODUCTION

No country has all the minerals it needs and no country would like be left behind in economic development. But the essential ingredient of economic development is the growth of basic industries and minerals provide vital raw materials for basic industries. Hence global trade in minerals is unavoidable. Thanks to ideological and technological revolutions, the world is but one market. The greater the competitiveness of any country, the bigger the share in the world trade.

The Indian economy is fast getting integrated with the world economy. The pace of transition is remarkable. The mineral sector cannot lag behind. With its vast natural resources our country is a serious competitor in the global market. Not that we are endowed with all the mineral resources but we have a vast geological potential. Already sixty-four minerals are being produced. Many more await discovery and economic exploitation. Out of these, large number of minerals are produced in small and medium scale mines which given due support promise to be highly competitive.

## GLOBAL MINERAL MARKET

Minerals contribute an important segment of world trade (685,000 million dollars in 1991). The share of minerals in the world trade was 19% in 1990. The Industrially advanced nations of Western Europe, North America and Japan are the main consumers of minerals in the world today. A good number of Asian countries are being rapidly industrialised and this is creating new demands for minerals. Some countries like Japan have mineral-based industries which depend almost entirely on imports, others like U.S.A. have to import large quantities of minerals, as their own domestic industry which is by no means small, is not adequate to meet their growing demand.

So far there has been cyclical rise and fall in global demand but it is expected that with faster industrialisation, smoother trade relation and easy transportation., there will be steady rise in demand and a greater stability in trade.

Reserves of important minerals in India and the world are stated in Annexure A1.

## THE INDIAN SCENARIO AND THE ROLE OF SMALL SCALE MINING SECTOR

At the moment Indian mineral trade is with 104 countries. The major countries to which minerals are exported are the USA, Japan, Hongkong, Belgium, Thailand and Italy. Out of 64 minerals currently being produced important minerals being exported are iron ore, dimension stones, chromite, mica, manganese, alumina, baryte, cut and polished diamond and precious stones. We are self-sufficient in 36 minerals but big deficits exist in fuel minerals, gold, copper, nickel, zinc, tin, lead tungsten, rock phosphate sulphur and raw diamonds. Unfortunately, import exceeds export. In 1994–95, minerals worth Rs. 27932 crore were imported while the export was worth Rs. 18587 crore only, leaving a deficit of Rs. 9345 crore. Minerals in India contribute 22% of the total export and 31% of the total import. Annexure A2 provides information regarding minerals resources in India under four categories viz. abundant, adequate, deficit and scarce.

Annexure 'B' indicates the production of 27 minerals during the last five years and Annexure 'C' shows the trend of major export and import. The value of mineral production during 1995–96 was Rs. 32,385 crores.

## ROLE OF SMALL SCALE MINE

Out of the total number of around 4400 mines operated in India, 85% can be classified in the segment of Small Scale Mining Sector. This sector accounts for 35% of total non-fueld minerals production. They are situated mostly in underdeveloped areas, quite a few operate in clusters and some of them are artisanal. A large number of them are tackling small mineral deposits of ores which may not be amenable to large scale operations. Such minerals are: apatite, asbestos andalucite, baryte, bauxite, bentonite, china clay, chromite, felspar fireclay, graphite, magnetite, mica, ochres, soap stone, quartz, garnet, varieties of rare earths, iron ore, manganese, zinc, fluorspar, phosphate, magnesite, granite, marble slate, limestone, etc. In view of the low tenor of ores and small size of some deposits, small scale mining is of great relevance. To make them cost-and-quality-competitive, induction of latest but flexible technology can be great value. At present, however, out-dated technology, higher costs of production and poorer quality of products are major handicaps.

## FACTORS WHICH INFLUENCE COMPETITION

As the majority of Indian mines are of small or medium size, their competitiveness will play a vital role in the mineral trade and business as a whole. But can they be competitive on their own ? To answer this question, let us examine the important factors which determine competitiveness. These factors are:

- Nature of Government Policies and strategies
- Level of development of infrastructure
- Domestic economic strength
- Domestic demand and rivalry
- Extent of internationalisation
- Scientific and technological capacities
- Quality of human resource and knowledge resource
- Quality of management cadre
- Institutional infrastructure
- Level of performance of capital markets and financial services.

The three requisites for maintenance of competitive edge are:

(a) superior quality
(b) lower price
(c) delighting services.

One of the classical examples of the products of small scale Indian mines succeeding in the world market has been mica. More than 90% of the world market was in Indian hands and mica came from small scale mines. Granites and marbles are other examples.

It is for the Government, the producers, and the supporting industries and institutions to play proper roles in the changed environment to maintain the competitive edge. Obviously, the strategy would consist of three-pronged endeavour to: (a) reduce import of minerals by indigenous production, (b) increase export to traditional markets and (c) expand market to new areas, keeping in mind the prerequisites stated above.

## WHAT ARE THE THROTTLES ?

The small and medium scale mining sector is throttled in many ways.

Take for instance, the short periods of mining lease and the threat of its premature termination embodied in the mining lease deed. The lack of security of tenure discourages long term adequate investment. At the level of the States which are the main locations of mining activity, implementation of policies has been lamentably tardy due to various bureaucratic and procedural hurdles. Legislative amendments to facilitate rapid progress have come in trickles and are far too little. The result is that the small mines derive hardly any benefit of liberalisation.

There is a plethora of mining laws and and surprisingly all laws are applicable equally, irrespective of whether a mine produces a few hundred tonnes or a million of tonnes per annum. Departmental guidelines are cumbersome and rigorous; the implementation of these laws and guidelines entail too disproportionate a financial burden. There are at least 19 types of taxes/levies, more than 20 Govt. departments having finger in pie and more than 30 Acts/Regulations which ensnare these mines.

Grant and renewal of mining leases are inordinately delayed; any change in the mine plan to suit the leasehold is objected to; royalty and dead rent are high and not stable for any reasonable periods.

The owners of small mining leases are called upon to prepare

- Mining Plan
- Environment Impacts Assessment Plan
- Environment Management Plan
- Proposals for clearances under Forest Conservation Act
- Biological Reclamation Plan, etc.

The procedures laid down are such that it takes years to obtain Govt. clearance. Besides, the financial burden in preparing plans and documents and going through the procedure is so high that the operators of small mines can seldom shoulder. These are serious harassments which are highly discouraging. On the top of it all, there are obstacles created by self-appointed environmentalists who make populist moves without weighing properly the insignificance or otherwise of the damage to environment and the activities carried out to enhance the carrying capacity of the eco-system and thereby achieving sustainability. The exaggerated enthusiasms to protect environment has led to the opposition to production and export of high quality granites from Chandardings hills of Assam.

Some State Governments have imposed tax at exorbitant rates on excavation and use of forest land for non-forest purpose.

There is serious lack of infrastructural facilities particularly, power, roads, transport system, communication, institutional financing, general know-how, testing ports etc. Take for instance, the transport of minerals to port. Although double axle taurus trucks are capable of transporting 24–25 M. tonnes of weights (carrying a net weight of 21.5 M.T. inside the container), due to limits imposed by Motor Vehicle Act, exporters are not allowed to stuff more than 16–19 tonnes in the container. As a result, cost of transportation goes up. Similarly, the ports nearest to mining areas have limited capacity and minerals have to be carried to ports at longer distances, making the business costlier. Again, for making a container, Indian ports need cash outlay of $ 520 while other ports in South Asia cost $ 300.

Loss of minerals during transport by railways is yet another malady which tells upon the overall business. Steel mills and power houses have suffered and would prefer imported coal particularly for coastal plants.

Small scale mining Sector is saddled with the same labour laws and safety laws as apply to the biggest of the industries. Labour laws do not allow exit of unwanted workers. Safety laws necessitate statutory number of supervisors, irrespective of the small number of man power employed. It also makes it mandatory to prepare and submit a large number of returns, proforma, reports, plans, etc.

## STEPS NEEDED TO IMPROVE COMPETITIVENESS

The objective is to encourage small mine to (a) cut down the cost, (b) improve quality, (c) facilitate export and provide due services. It is well known that something extra in product and sevice provides decided advantage over the competitors. To meet the objective certain calculated steps have to be taken.

National Mineral Policy should be suitably amended to provide special dispensation to small mineral deposits and small scale mines.

The legislations need amendment so as to relieve small mines from certain provisions which apply to large mines, provisions regarding areas of lease-hold, tenure of lease, preparation of plans and documents, applicability of various taxes, etc. There should be distinction between operations of large and small mines. Renewal of mining lease for small deposits should be automatic till reserves are completely exhausted. Setting up of Mining Financial Corporations on the line of I.D.B.I. or State Finance Corporation should be expedited to facilitate availability of finance. Lack of suitable guidelines for foreign investment in mining sector is a deterrent. Small and medium scale mines attract foreign investment but because of uncertainty in the condition of investment in the country, it is not materialising.

There are other deterrents to flow of capital and technology. Mining is financially a high risk venture because of a number of unpredictable features. Before commencing actual mining operations one has to carry out two essential high risk activities. There are: (a) geological exploration and (b) mine development. An entrepreneur has to depend on his own equity and un-secured loan. other funding options are not easily available. The term lending institutions generally provide for financing equipment only and are reluctant to cover mine development. Recently, the World Bank is reported to be taking interest in small mines, particularly, those engaged in 'cluster mining'. This is a welcome move.

While the Govt. is providing incentives to small scale industries with highly beneficial results, the same has not been extended to small mines. These incentives can be in way of sales tax holiday, income tax exemption, waival of local taxes like octroi, financial subsidies, seed capital, etc. which apply to backward districts and far-flung areas. After all, small mines are also normally situated in comparatively backward and far-flung areas.

A comprehensive mining plan which must be made to ensure scientific mining should include Environment Management Plan which itself could be part of the regional environment planning prepared by a multi-disciplinary body. And for clearance under Forest Conservation Act. I.B.M. can be the final authority to approve both mining and environment plan including compensatory afforestation where needed.

Small scale mines should be exempted from extraneous taxes imposed by State Govt. such as tax on restoration and improvement of degraded land or land tax on excavation and use of forest land.

The concept of 'cluster mining' and 'rural mining' has inbuilt advantage of overall cost reduction and quality control.

The above measures will not only facilitate smoother operation but will also contribute substantially to cost reduction which is one of the main criteria for competitive edge.

So far as quality improvement is concerned, serious considerations and efforts have to be made.

Geological reserves of small deposits to which small scale mining is amenable, are not necessarily of high quality. For instance, deposits of minerals like coal, limestone, bauxite, iron ore, kyanite, graphite, magnetite and phosphate have poor quality. While mining operation has to be such that external impurities are not allowed to be mixed, great care has to be taken to prepare run of mine by sizing and beneficiating to transform it into superior quality raw material. This process involves right skill and additional financial support

## EXPORT POSSIBILITIES OF COAL AND OTHER MINERALS FROM SMALL AND MEDIUM COAL MINES

There are several coal reserves in the country which are patchy in nature and are situated away from the major coal fields. Several illegal small mines are already operating in clandestine manner in different parts of the country. Coal mining is predominantly in public sector but with liberalisation it is expected that small areas will be leased out to private companies after making suitable legislative amendment so that there is no incentive for any one to resort to illegal operation.

We have coal-starved neighbours such as Bangladesh, Pakistan and Nepal. Small coal mines of India will positively have edge over non-coking coal from abroad.

Besides, virgin area can be leased out to private parties who can beneficiate and cater to the need of domestic consumers and those of Asia Pacific Countries. Similarly, iron ore from Bihar, Orissa, Goa and Madhya Pradesh, Mica from Bihar, Andhra Pradesh and Rajasthan, magnetite and quartzite from Bihar and Orissa, chromite from Manipur, Sillimanite from Meghalaya, copper, lead, zinc and coal of low grade from

Sikkim, rare earths from Tamil Nadu, Kerala Gujarat can rightfully have a share in the export market.

It has been earlier suggested that for the purpose of giving support to the small/medium scale sector in the field of transporting, process development, beneficiation and other R&D studies and development work, a special fund should be created in every State with a certain percentage of income from royalty and cess and matching grant from the Central Govt. This arrangement will substantially increase revenue and other benefits to the economy and more than compensate the financial outlay.

The small scale mining sector has to generate capacity to absorb fluctuations in the market and face the strategies of competitors. Granite is a typical example. The industry is reported to be in tight spot, firstly, because the international price of titles and slabs has gone down and secondly because there is increased competition with Italy. Italy has accumulated huge stock of rough block for some years and is in commanding position regarding pricing.

The third crucial aspect of competition is the service to the customer. Mica mines in small sector had for years shown that adhering to time schedules of delivery and honouring commitments made in the business deal, extending courteous behaviour to the customers and taking due interest to keep them delighted paid dividend.

In this connection, bilateral and multi-lateral technical assistance agreements and encouraging joint ventures with foreign companies of matching size are likely to provide wider knowledge, newer technologies and access to the world market. Transfer of New Gravitational Gold Concentration Technologies to small mining sector of Atacama Region of Chile promises to raise productivity and provide solution to the problems of population. Indian entrepreneurs can also set up joint venture in small and medium sector with foreign countries. Namibia, for instance, has invited Indian business men to set up joint venture in small and medium mining sector. Yet another example of desirable collaboration is Australian keenness to participate in the field of mining. The Australian High Commissioner once described Orissa as Western Australia as the geology is very similar in these two parts. It is well known that Western Australia is the major mineral province of Australia and has the same range minerals as Orissa.

This effort, however, has to be supplemented by revamping Indian ports and Cargo handling which, unfortunately, are suffering at the moment from crises of: (i) capacity (ii) cost and (iii) delay, eroding competitiveness to a great extent.

The importance of human resources development to cope with the transformation cannot be overemphasised. Not only entrepreneurship and mine operation need better skills but the nitty-gritty of global marketing has also to be mastered.

The pursuit of economic diplomacy, clearing wrong impressions about quality of Indian minerals and selling India abroad need to be followed vigorously.

A large number of small and medium scale mines operate under State Govt.'s Mineral Development Corporations. These Corporations have to be allowed to run on business lines, and the Govt. must distance itself to avoid interference. Organisations have to be re-structured and unnecessary shackles of bureaucracy removed. Facilities of institutional support from research stations, testing laboratories, dissemination of information, co-ordination with regional planning agencies, national and global market intelligence should be made available. For metalliferous mines keeping to be informed about the price trend as indicated by L.M.E. (London Metal Exchange) is highly desirable.

Exploration of new market has to be continuous. This applies to domestic as well as international market. For instance, the domestic market for granite is on the rise because of boom in construction sector. Similarly, granite for monuments have a good demand in the world and with the wage content being high, Indian suppliers can have a price advantage of around 25%. In tiles and slabs we stand to lose in competition. Newly developed countries hold out great promises regarding new market.

## CONCLUSION

Given a reasonable entrepreneural freedom and assistance, the small scale mining sector will be in a position to maintain low cost of production, high quality and delighting services to customers — the vital factors for competitive edge. Our small mines of iron ore, dimension stones, gem stones, mica, chromite and bauxite have given eloquent proof of their capabilities. Indian entrepreneurs have proved their merit abroad as well. With congenial industrial climate, small and medium mines can come up to expectations.

## Annexure A1
## MINERALS RESOURCES AVAILABILITY IN INDIA

| Sl. No. | Mineral | Unit | Reserves in India as on 1.4.90 | Reserves in World |
|---|---|---|---|---|
| 1 | Antimony | '000 tonnes | 11 | 4,700 |
| 2 | Apatit and Rock Phosphate | '000 tonnes | 115 | 40,000 |
| 3 | Asbestos | '000 tonnes | 2,295 | 104,000,000 |
| 4 | Barytes | '000 tonnes | 70,147 | 500,000 |
| 5 | Bauxite | '000 tonnes | 2,252 | 28,000,000 |
| 6 | Bentonite | '000 tonnes | 367,557 | 1,600,000 |
| 7 | Chromite | '000 tonnes | 88,351 | 6,800,000 |
| 8 | Coal & Liginite | '000 tonnes | 192,359,150 | 7.5–14 (million tonnes) |
| 9 | Cobalt Ore | '000 metric tonnes of cobalt content | Practically Nil | 8,800,000 |
| 10 | Copper Ore | '000 tonnes | 324,792 | 590 (m.t. of Cu.content) |
| 11 | Dolomite | '000 tonnes | 4,967,467 | — |
| 12 | Fluorspar | '000 tonnes | 2,149 | 310,000 |
| 13 | Felspar | '000 tonnes | 16,138 | 1,000,000 |
| 14 | Fireclay | '000 tonnes | 696.72 | 76,000 |
| 15 | Fuller's earth | '000 tonnes | 228,300 | 4,900,000 |
| 16 | Gold | tonnes of metal | 101 | 51,000 |
| 17 | Gypsum | '000 tonnes | 239,312 | Large |
| 18 | Graphite (natural) | '000 tonnes | 3,109 | 380,000 |
| 19 | Ilmenite | '000 tonnes | 86,750 | @ 430,000 |
| 20 | Iron Ore | '000 tonnes | 12,745.00 | 230,000,000 |
| 21 | Kyanite | '000 tonnes | 2.715 | Large |
| 22 | Kaolin | '000 tonnes | 985,969 | 13,000,000 |

Contd.

**Annex A1 contd.**

| | | | |
|---|---|---|---|
| 23 | Lead & Zinc Ore | '000 tonnes | 167,555 | 130,000+330,000=460,000 |
| 24 | Limestone | '000 tonnes | 76,446,006 | Adequate |
| 25 | Magnesite | '000 tonnes | 233,328 | * 3,400,00 |
| 26 | Molybdenum Ore | '000 tonnes | 8,037 | * 12 |
| 27 | Manganese Ore | '000 tonnes | 176,477 | 4,800,00 |
| 28 | Mica | '000 tonnes | N.A. but large | large |
| 29 | Nickle Ore | '000 tonnes | 294,000 | *110,000 |
| 30 | Petroleum oil (crude) | '000 tonnes | 757,400 | 455 (b. barrels) |
| 31 | Potash | '000 tonnes | 739,000 | # 17,000,000 |
| 32 | Pyrites | '000 tonnes | 91,520 | 2,700 |
| 33 | Quartz/Quartzite/Mica sand | '000 tonnes | 1,349,000 | N.A. |
| 34 | Rutile | '000 tonnes | 5,180 | 85,000 |
| 35 | Salt (rock) | '000 tonnes | limited | large |
| 36 | Talc/Soapstone & Steatite | '000 tonnes | 83,665 | large |
| 37 | Titanium | '000 tonnes conc. | 14,000 | *700,000 |
| 38 | Tin Ore | '000 tonnes | 28,907 | *10,000 |
| 39 | Tungsten Ore | '000 tonnes | 23,894 | *3,400 |
| 40 | Thorium | '000 tonnes | 550 | ## 1,400,000 |
| 41 | Uranium | '000 short tonnes | 61 | 12,175 |
| 42 | Vanadium Ore | '000 tonnes | 13,337 | * 27,000 |
| 43 | Vermiculite | '000 tonnes | 313 | 200,000 |
| 44 | Wollastonite | '000 tonnes | 4,286 | N.A |
| 45 | Zirconium | '000 tonnes | 1,200 | 58,000 |

Source: US Bureau of Mines, Washington; Indian Bureau of Mines, Nagpur
Non-Fuel Minerals and Foreign Policy Royal Institute of International Affairs, London

Note
* Metal content
# $K_2O$ equivalent
## $ThO_2$ content
@ $TiO_2$ content
** $ZrO_2$ content

## Annexure A2
## MINERAL RESOURCES AVAILABILITY IN INDIA

| Grouping | Abundant | Adequate | Deficit | Scarce |
|---|---|---|---|---|
| Fuel Minerals | Coal (Non-coking) | — | Coal (coking) Petrolieum Crude oil | — |
| Metallic Minerals Ferrous | Iron Ore | Chromite (Metallurgical grade), Manganese Ore | Chromite (Refractory grade) | Nickel Tungsten, Cobalt, Moylbdenum, Vanadium |
| Non-ferrous | Bauxite (Metallurgical grade) | Zinc | Bauxite (Chem, Refractory grade), Copper, Lead | Antimony, Gold Platinum group of metals, Silver, Tin |
| Industrial Minerals | Clay, Barytes, Ball Bentonite, Calcite, Dolomite (excluding low silica grade), felspar, fireclay, Fuller's Earth, Grante, Gypsum., Kaolin, Limestone, Magnesite, Mica Pyrophyllite, Pyrite, Ochre, Quartz, and Silica sand, Quartzite, Sillimanite, Zircon, Steatite | Carborandum Graphite, Rock Salt, Vermiculite, Wollastonite | Apatite Rock, Phosphate, Asbestos, Fluorite, Kyanite | Andalusite, Borax Diatomite Native Sulphur, Potash |
| Precious stones | — | — | — | Diamond, Emerald, |
| Other Minerals | Granite (Dimension stone), Marble | — | — | — |

Source: C.M.R.I.

## Annexure B
## PRODUCTION OF VARIOUS MINERALS (INDIA)

| Sl. No. | Name of the minerals | Unit | 1990–91 | 1991–92 | 1992–93 | 1993–94 | 1994–95 |
|---|---|---|---|---|---|---|---|
| 1 | Barytes | Th. tonnes | 509 | 684 | 405 | 617 | 444 |
| 2 | Bauxit | Th. tonnes | 4984 | 5013 | 5121 | 5450 | 4855 |
| 3 | Chromit | Th. tonnes | 940 | 1082 | 1071 | 1055 | 1049 |
| 4 | Dolomite | Th. tonnes | 2648 | 2932 | 3051 | 3509 | 3117 |
| 5 | Felspar | Tonnes | 73863 | 69420 | 74086 | 71199 | 63528 |
| 6 | Fireclay | Th. tonnes | 536 | 531 | 462 | 418 | 397 |
| 7 | Graphite | Tonnes | 63953 | 77384 | 74679 | 76214 | 93696 |
| 8 | Gypsum | Th. tonnes | 1589 | 1582 | 1802 | 1689 | 1539 |
| 9 | Iron Ore and Concentrates | Th. tonnes | 55591 | 58834 | 57147 | 58338 | 58367 |
| 10 | Kaolin | Th. tonnes | 724 | 799 | 668 | 656 | 655 |
| 11 | Limstone | Th. tonnes | 70125 | 77185 | 76617 | 83705 | 86194 |
| 12 | Magnesite | Th. tonnes | 529 | 531 | 541 | 374 | 336 |
| 13 | Manganese Ore | Th. tonnes | 1429 | 1640 | 1903 | 1677 | 1632 |
| 14 | Mica: | | | | | | |
|  | Crude | Tonnes | 4,602 | 3,593 | 2,560 | 2,086 | 2,041 |
|  | Waste & scrap | Tonnes | 3,366 | 2,364 | 1,758 | 1,019 | 717 |
| 15 | Orche | Th. tonnes | 120 | 126 | 139 | 144 | 164 |
| 16 | Quartz | Th. tonnes | 216 | 192 | 205 | 160 | 133 |
| 17 | Silica Sand | Th. tonnes | 1457 | 1497 | 1208 | 1321 | 1181 |
| 18 | Steatite | Th. tonnes | 431 | 452 | 433 | 401 | 338 |
| 19 | Wollastonite | Tonnes | 59722 | 62493 | 54183 | 93776 | 71305 |
| 20 | Vermiculite | Tonnes | 1806 | 1651 | 1455 | 2319 | 1899 |
|  |  |  | 1988–89 | 1989–90 | 1990–91 | 1991–92 | 1992–93 |
|  | **Minor Minerals** | | | | | | |
| 21 | Bentonite | Tonnes | 225,953 | 246,362 | 174,682 | — | — |
| 22 | Fuller's Earth | Tonnes | 25748 | 35325 | 20331 | — | — |
| 23 | Granite | Cu. metres | — | 341720 | 277129 | 116378 | — |
| 24 | Marble | Cu. metres | — | 515701 | 686123 | 789183 | — |
| 25 | Slate | Tonnes | — | 30591 | 25050 | 21273 | — |
|  | **Atomic minerals** | | | | | | |
| 26 | Ilmenite | Tonnes | 237,143 | 256,835 | 300,525 | 258,769 | — |
| 27 | Rutile | Tonnes | 10475 | 9595 | 13016 | 13151 | — |

Source: Indian Bureau of Mines, Nagpur

## Annexure C
### EXPORTS AND IMPORTS OF SELECTED MINERALS METALS AND MINERAL BASED PRODUCTS, 1990–91 TO 1992–93

| Item | 1990–91 | | | 1991–92 | | | 1992–93 | | |
|---|---|---|---|---|---|---|---|---|---|
| | Export | Import | Net | Export | Import | Net | Export | Import | Net |
| **Nonferrous Metals & Alloys** | 379.7 | 1354.0 | 974.3 | 467.1 | 1105.5 | -638.4 | 704.8 | 1393.0 | -688.2 |
| **Fuel Mineral** | 1012.9 | 11516.1 | -10503.2 | 1230.2 | 140065 | -12834.8 | 1656.8 | 18318 | -16661.2 |
| Coal | 8.7 | 756.4 | -747.7 | 15.2 | 0.7 | -885.5 | 50.2 | 1309.4 | -1259.2 |
| Coke | 0.3 | 32.9 | -32.6 | 0.3 | 135.8 | 135.5 | 0.3 | 73.6 | 73.3 |
| Petroleum(crude) | 0 | 6066.6 | -6066.6 | 0 | 7810.2 | -7810.2 | 0 | 10575.4 | 10575.4 |
| Proroleum product (POL) | 1003.9 | 4660.2 | -3656.3 | 1214.7 | 5218.3 | 13.6 | 1606.3 | 6359.6 | -4753.3 |
| **Ferrous Minerals** | 1,135 | 78.8 | 1056.2 | 1566.5 | 99.6 | 1466.9 | 1214.4 | 48.1 | 1166.3 |
| Chromite | 41.7 | 0.1 | 41.6 | 93.6 | 0 | 93.6 | 78.9 | 0.2 | 78.7 |
| Iron ore incld. concentrates | 0 | 64.7 | 64.7 | 0 | 85.1 | 85.1 | 0.6 | 30.9 | 30.3 |
| Manganese Ore | 44.2 | 2.5 | 41.7 | 37.5 | 3.4 | 34.4 | 308 | 3.7 | 27.1 |
| Nickel Ore & concentrate | 0 | 64.7 | -64.7 | 0 | 85.1 | -85.1 | 0.6 | 30.9 | -30.9 |
| **Ferrous Metals & Alloys** | 442.5 | 3011.5 | -2569 | 787.1 | 2533.6 | -1746.5 | 1280.6 | 3521.2 | -2240.6 |
| Charge Chrome | 52.4 | 0 | 52.4 | 134.9 | 0 | 134.9 | 175.5 | 0 | 175.5 |
| Ferro Chrome | 5.2 | 0.3 | 4 | 22.3 | 0.8 | 21.5 | 3.6 | 2.5 | 33.1 |
| Ferro Manganese | 14.8 | 0.7 | 14.1 | 11.5 | 0.2 | 11.3 | 14.9 | 0.2 | 14.7 |
| Ferro Alloys (others) | 3 | 109.2 | -106.2 | 8.7 | 135.8 | -127.1 | 25.4 | 128.7 | -103.3 |
| Iron & Steel (Material) | 362.3 | 1867.9 | -1505.6 | 605.1 | 1693.2 | -1088.1 | 8 | 131 | -904.2 |
| Iron & Steel (scrap) | 4.8 | 1033.4 | -1028.6 | 4.6 | 703.6 | -699 | 2.4 | 1458.8 | 4456.4 |
| **Non-ferrous minerals** | 13.4 | 67.3 | -48.9 | 0 | 27.3 | -23.3 | 25.3 | 113.7 | -88.8 |
| Lead Ores & concentrates | 0 | 25.5 | -25.5 | 0 | 10.8 | -10.8 | 0 | 42.1 | -42.1 |
| Lead Ores & concentrates | 0 | 41.2 | -41.2 | 0 | 16.4 | -16.4 | 0 | 71.2 | -71.2 |
| Copper Ore & concentrate | 13.7 | 0.1 | 13.6 | 0 | 0 | 0 | 0 | 0.3 | -.03 |
| Bauxite | 4.7 | 0.5 | 4.2 | 4 | 0.1 | 3.9 | 25.3 | 0 | 25.2 |
| Alumina | 199.2 | 11.1 | 181.1 | 142.8 | 110.2 | 132.8 | 162 | 15 | 146.5 |
| Aluminium | 155.2 | 121.3 | 33.9 | 300 | 103.8 | 107 | 480.2 | 82 | 398.2 |
| Copper and alloys incld. brass and bronze | 20.8 | 818.4 | -797.6 | 15.1 | 724.3 | -709.2 | 28.9 | 96.5 | -967.6 |

(Contd.)

(Annexure C Contd.)

| | | | | | | | | | |
|---|---|---|---|---|---|---|---|---|---|
| Lead and alloys includ. scrap | 1.5 | 67.8 | -66.3 | 2.3 | 38.7 | -36.4 | 4.4 | 20.8 | -16.4 |
| Nickel & alloys includ. scrap | 0.4 | 79.8 | 79.4 | 3.5 | 78.6 | -75.4 | 15.4 | 144 | -128.6 |
| Silver | 0.4 | 6.8 | -6.4 | 0.2 | 13.5 | -13.3 | 0.3 | 17.8 | -7.5 |
| Tin & alloys includ. scrap | 1.7 | 26 | -24.3 | 2.1 | 21.9 | -19.8 | 8.1 | 29.8 | -21.7 |
| Tungsten & alloys includ. scrap | 0.3 | 2.8 | -2.5 | 0.2 | 6.3 | -6.1 | 0.5 | 8.4 | -7.9 |
| Zinc alloys includ. scrap | 0.2 | 220 | -219.8 | 0.4 | 108.4 | -1098.3 | 5 | 78.2 | -73.2 |
| **Chemical & Fertilizer** | 1.4 | 1578.5 | -1577.1 | 7 | 2879.6 | -2872.6 | 7.2 | 3068 | -967.6 |
| Fluorspar | 0 | 11.6 | -11.6 | 0.1 | 10.3 | -10.2 | 0 | 6.7 | -6.7 |
| Phosphoric Acid | 1.4 | 498.9 | -497.5 | 6.6 | 1540.5 | -1533.9 | 6.8 | 1676 | -1669.2 |
| Potash (fertilizer) | 0 | 443.3 | -443.3 | 0.2 | 567.1 | 566.9 | 0.3 | 575.4 | -575.1 |
| Rock Phosphates | 0 | 346.4 | -346.1 | 0 | 454 | -454 | 0 | 458.4 | -458.4 |
| Sulphur | 0 | 278.3 | -278.3 | 0.1 | 307.1 | -307 | 0.1 | 351.5 | -351.4 |
| **Prec. & Semi Prec. Stones** | 4908.9 | 3721.5 | 1187.4 | 6039.4 | 4810.2 | 1229.2 | 8063.2 | 7059.6 | 1003.6 |
| Diamond | 4711.8 | 3599.6 | 1112.2 | 5761.8 | 4695.8 | 1066 | 7785.2 | 6892.9 | 892.3 |
| Emerald | 100.8 | 90.4 | 10.4 | 153.6 | 74.1 | 79.5 | 157.7 | 102.1 | 55.6 |
| Others | — | 31.5 | 64.8 | 124 | 40.3 | 83.1 | 120.3 | 4.6 | 55.7 |
| **Industrial Minerals & others** | 2.4 | 138.4 | 136.7 | 468.8 | 172.3 | 296.3 | 633.8 | 40.3 | 43.5 |
| Magnesite | 1.8 | 46.1 | 44.2 | 2.1 | 64.6 | -62.5 | 1.2 | 85.7 | 84.5 |
| Granite | 198.1 | 0.4 | 197.7 | 373.8 | 2.9 | 370.9 | 492.8 | 0.2 | 492.6 |
| Marble | 2.5 | 1 | 1.5 | 13.7 | 2 | 1.7 | 27.3 | 3.7 | 33.6 |
| Asbestos | 0 | 72.5 | -72.5 | 0 | 84.9 | -84.9 | 0.1 | 86.1 | 86 |
| Limestone | 1.6 | 15.3 | -13.7 | 1.9 | 14.9 | -13 | 1.8 | 2.8.9 | -27.1 |
| Mica | 51.3 | 2.9 | 48.4 | 55.5 | 2.8 | 52.7 | 38.3 | 5.7 | 32.6 |
| Portland Cement | 19.7 | 0.2 | 19.5 | 21.8 | 0.2 | 21.6 | 62.3 | 0 | 62.3 |
| **Total** | 8173.9 | 21466.1 | -13292.2 | 10570.1 | 25693.1 | -15123 | 13586.1 | 33731.9 | -20145.8 |

Source: FIMI

# ECONOMICS OF SMALL SCALE RESOURCES: A COLLECTIVE MINING EFFORT

[1] *Suman Banerjee,* [2] *Sayeri Banerjee,* [3] *Sajal Dasgupta and* [4] *S.M. Chatterjee*

[1] B.E. College, [2] Dept. of Economics, Calcutta University, [3] Head, Dept. of Mining Engineering, B.E. College, [4] Director, B.E. College

## INTRODUCTION

Small scale mining has earned a dubious entity in terms of people centered state owned, employment generating sector and on the other hand, a threat to the environmental protection agencies. Despite being a very small segment it is conceptually and technically a paradigm shift from the traditional mining and provide an organised economical means in the mineral sector.

Despite having a controversy with regard to it's definition, i.e., how small an operation could be that can be universally recognised as small scale mining, it more or less refer to sporadically occurring, low volume multi/single mineral deposits and its method of extraction.

The significant advantages of small scale mining in our country are as follows:

(i) Exploratory activities can give rise to production
(ii) Environmental damage can be easily restricted
(iii) Small scale operations—may be even family controlled, a rural version of SOHO type of organisation.
(iv) Being labour intensive, it can definitely generate local employment.

Small mining, decidedly plays an important role in the world production of minerals and contributes significantly to the national economy in case of many countries (e.g. Kuroko deposit of Japan). In national mineral policy of India, a special discretionary effort has been emphasized to promote small scale mining in a scientific and efficient manner while safeguarding the vital environmental and ecological imperatives. The present paper deals with the economical aspects of small scale mining.

## INSTABILITY IN MINERAL MARKET

The instability in the prospect of mineral industry is mainly due to a major shift in the business dynamics in mineral sector. The growth of any industry is related to its market. Investment is now getting adopted on the long term to consumption growth and on the short term, supply adjusted to demand. This can be referred to as the end of Oligopolies. The main reason for an oligopolistic growth in the mineral market was:

(i) Relatively regular growth of mineral and metal consumption.
(ii) Existence of differential rents related to the specificity of the deposit and is carefully nurtured by Government policies.
(iii) The capacity of certain major producers to control prices through supply adjustments on the market.

An oligopolistic structure does not strictly rely on supply concentration. It hinges on a balance between investment and growth demand through market price control. The price leaders effectually controls the balance through a consensus between producer and consumer. For a price leader, a temporary decrease in production due to low demand has become useless to generate a price hike. It is tantamount to speaking in terms of market economy that the entire mineral sector has dropped down from the position of "Cash Cow" to "Dog" category. As the inertia of growth demand is increasingly falling and demand is globally unpredictable, investment in new capacities even applying to low cost operations, appears to be more risky than the past. This uncertainty is not going to be changed unless the supply-demand adjustment resurrects.

## SCOPE OF SMALL SCALE MINING

The most basic problem in context with the small scale resource is its incompleteness in information about the properties of mineral deposits. Deposit size and shape are related to the tonnage of the deposit and may affect how a deposit is mined. Deposit orientation affects the choice by which the deposit can be mined. The spatial distribution of deposits is important for resource assessment, search strategies and costs of exploitation. Apart from the above, two derived properties of deposits are metal grade and tonnage. A slight shift in the mean of low grade can prove a deposit worth exploitable which is particularly important in case of low grade sporadic resources. It also demands a different approach from the supply logistics which has been discussed in the subsequent sections. Optimum physical conditions for the successful operation of small scale mines are essentially the same as those most helpful for large operation. However, conditions that minimize initial development costs, surface facilities and capital equipment are essential for small operation. Due to having scant reserves, a limited total financial return has to amortize the above cost. Development expenses fo small underground mines are minimised when mining can commence from outcrop. Generally, the costs for deep shaft or shaft sinking are prohibitive. Though illegal, "rat hole" mining in many parts of West Bengal are examples of improvisation of local knowledge to substitute costly ventures.

## COLLECTIVE MINING

Unless Giffin's Paradox holds good, the market situation in mineral industry is not going to change. One of the most effective way to safeguard the interest of small mining sector is to provide a different organisational structure far apart from traditional mining organisations. This was proposed by the Russians that mines can be classified into three categories according to organisational and technological conveniences:

(i) Self-contained mine
(ii) United mine
(iii) Central mine

Collective mining essentially borrows the concept of central mine of U.S.S.R which comprises several mines as separate entity but connected to central facility through transport, processing and maintenance facility. Total cost component of a mine consists of fixed and variable cost, and the fixed component is straightway proportional to degree of mechanisation and complexity of deposit, the best way to combat the situation is to divide the fixed cost and minimize the recurring cost through central facility. Collective mining concept is economically viable and can be organised through local/state Government.

## ROLE OF GOVERNMENT

Government should not only organize the central facility but also patronize the marketing of products from small scale mines. Mined materials are not always directly saleable in the market. A small mine rarely has the reserves to provide a mine life sufficient to justify investment in further processing facilities. Consequently, it is the aim of small mine to avoid further processing. Government should provide a satisfactory concrete marketing agreement which will transfer the risks of further processing elsewhere, yet assure payment for the contained metals at market prices.

## CONCLUSION

The paper above hinted a direction to make small scale mining a profitable venture economically. However, its implementation requires lot of

Government intervention and help. The small size results in a disadvantage with respect to working capital and tendency towards unequal pricing. As mentioned at the outset, the lack of domestic facilities and uncertainties of transport should not hamper the encouragement of small mines and with the generous support from the Government, it is possible to make the venture profitable.

# SECTION 6

# Miscellaneous Papers

# AN INSIGHT INTO ENTREPRENEUR–EMPLOYEE RELATIONSHIP IN SMALL-SCALE MINING ACTIVITIES

S.K. Mukhopadhyay[1] and J. Bhattacharya[2]
Deptt. of Mining Engineering, IIT, Kharagpur 721302, India

## INTRODUCTION

The level of industrial growth to a large extent can be attributed to the development of mineral resources. India is fortunately endowed with resources of many important minerals, distributed over numerous large and small deposits located all over the country. The number of reporting mines excluding minor minerals and atomic minerals totalled about 4400 towards the end of the last decade, contributing to a value of about Rs. 16.7 thousand crores, employing about 1.05 (IBM Yearbook, 1991) million persons. According to one computation, out of 60 minerals mined in India 38 of them come from small-mines contributing to around 20% (by value) of the total mine output—about 6% coming from 'Minor-Minerals' and 14% from Non-Minor Minerals.(SSMI, 1984).

## FEATURES OF SMALL-SCALE MINES

Mines operating in large mineral deposits with large-scale operations generally have forward integration with industrial production units. These large deposits are getting depleted at a faster rate owing to high production and large-scale mechanization. To cope up with the demand-supply balance, small deposits having small-scale mines are slowly receiving importance. Although, the majority of main mining activities have been developed out of small mining operations, the economic and social contributions of small-scale mining have long been overlooked.

---

[1] Associated Professor
[2] Assistant Professor

## Techno-economic Features

Small-scale mining in most cases are rarely the result of systematic planning but by and large a product of chance. A small-mine is opened in the event of deposits becoming visible through conspicuous oxidation of outcrop and minerals being noticeable through natural enrichment. These characteristics of small deposits lead to straightforward mining in a small-scale.

The small-mines account for approximately 16% of the world-wide output of mineral products. Like 'ores', 'metal compounds' and 'Metals' excavated from small-mines, the 'industrial minerals' like quartz, sulphur, graphite, fluorspar, kaolin, gypsum, carbon, mica, diamond, semi-precious stones on the other hand are also excavated from small mines, which occur in a comparatively pure and non-intergrown form. They have high market value and can be extracted from the deposits using simple methods and easily processed into marketable products. The isolated locations of the small scale mines compel the miners to process the raw ore on the mines premises itself to obtain commercial concentrates. Possibility of custom milling and dressing are rare. Even if the contribution of individual small mining activities is little, together they form a formidable product range and a good export earner. It is reported that in Bolivia, over 40% of countries mineral export is contributed by small-scale mining products. Small-scale mining have in common; (Mukhopadhyay and Ramlu, 1985).

— very low level of mechanization;
— low capacity;
— labour intensive techniques; and
— low capital intensive operations.

Small-mines possess, in general, a higher real net product in proportion to investment of capital and are able to go into production more quickly. They have low capital lockup and are better equipped to adjust to changing market conditions. Continuation of exploitation of small deposits stimulates further regional prospecting and exploration possibilities.

## Socio-economic Features

In regard to industrial status, small-scale mining is still considered as an unorganized industrial sector which, from both local and central governments, alike receive a step-motherly treatment. Contractual and license raj still reigns supreme. Located in far-flung areas, isolated, disadvantaged by lack of power and infrastructure small-scale mines have always been considered as small-time investment and never have been considered as a continuous stream of income generation, both during and after mining. Unfortunately, viewed as a quick money making proposition before leap frogging into a new locale small-scale mining is ill-supervised, neglected and to sum it all, condemned. But its respects are due.

Small-scale mining contributes towards the improvement in the social-sphere around its locale, provided this sector is given some attention in the interest of the state and the workforce engaged in operations. It contributes towards development of rural areas. Small mining operations help to create improved infrastructural facilities, like approach roads energy and water supply in a low-scale. Small mining activities, being situated in economically backward areas, stimulate greater variety in income distribution and create new job positions. Consequently, the social and economic standard of living of the people in place, starts rising further. Apparent contribution to government are the taxes and royalties from operators. But a significant contribution is fulfilling the social planning, the objectives of rural income generation, infrastructural development and stoppage of migration of rural labour.

## ATTITUDE: ENTREPRENEUR VIS-A-VIS EMPLOYEE

Small-scale mining is predominantly labour intensive. The operators are mostly local residents, comprising of unskilled individuals, families or family-groups. As they are unskilled, uneducated, cheap and available in plenty, historically many small mining operators adopted a pattern of employment; catch them young, force them to overwork even if necessary by small incentives, fire them whenever needed without compensation and again recruit new people. All these resulted in poor skill, safety of operations and flouting of prevailing laws. But with the call of social justice even rising throughout the country, it is time for both the operators and law enforcement agencies alike to understand that good employment practices are beneficial for the all-round development.

Therefore, besides pecuniary aspects, emphasis on human efficiency is one of the most pivotal factors behind sustenance of such small-mines. The workforce sometimes work in conditions having potential risk of accident, poor health service, low wage and delayed payment. A tendency persists among deposit developers in such a sector, who 'just want to conduct their business' without any participation in any actions that have overtones of employees or workforce. Such a case is sighted later in this article. Though, it is true that small-scale mine brings rapid return to the investment, some projects get plagued by start-up and running problems due to the above selfish attitude of the entrepreneurs. The workforce on the other hand, lack in interest to put their best effort. The resultant outcomes are delay, time-overrun and low productivity, which get reflected on company's profitability statement and on unit cost of production. Low productivity necessitates a higher number of employees for the same quantum of work, adding a new dimension to personnel problems.

## 'WIN/WIN' RELATIONSHIP—A SYNTHESIS

The problems discussed above, however can be solved by establishing 'win/wing' relationship between the entrepreneur and the employees. (McAlpine, 1992) This provides both parties equipollently, the opportunity to 'share and care' mutually, to ensure bright furture of the organization. Executives/entrepreneurs/investors/developers of small-mines must realize that the future of the organization largely depends on the performance of its employees and not on 'cost-cutting measures' alone. It must be remembered that the mainstay of small-scale mining is the human workforce and these human resources are the most valuable resources besides their acquired mineral resources for accumulating profit. The labour productivity gets affected by low measures, health and welfare amenities. A high labour productivity will enable a small mining company to function effectively and harvest rich dividends for both the mine-owners and the employees.

## FORWARD STEPS

Small mining entrepreneurs must understand and should not be apprehensive about the involvement and value of the employees. Policies towards the employees should neither be unreasonable for unrealistic. The executives should always be in quest of identifying innovative ways and alternatives to incorporate employees in various but real business opportunities to avoid quandary, if so arises. This approach will culminate into a positive relationship between employers and employees.

Policies should be framed to see that all employees become eligible for 'more earning' as their 'skill on-the job' progress. A major factor here is to eliminate the equanimity of keeping 'waiting' or allowing for 'demand to come' from the employees. This directly affects the human efficiency due to lack of motivation. A panacea for all such 'industrial-ills' is to call for the 'win/win' relationship, a change in 'management perspective'.

For inculcating 'win/win' relationship both the employees and employers as well as the state, to whom the property belongs, have to have a long term commitment. A recent amendment of MMRD Act-1958 has given the State Government more power regarding the issues on terms and conditions of the mining lease.

### Entrepreneurial Actions

The entrepreneurs must remember that

— employees/workforce are the most valuable resources and their involvement in management is a fact and that should not be ignored.

- the valuable suggestions given by the employees should not be underestimated;
- long term contracts for employment allows one to think the organization as one's own;
- start-up of incentive and reward schemes as a token of recognition, based on output and performance, help to motivate the workforce;
- the employees should be trained at all levels;
- proper tools are to be provided to the workforce, to carry out their jobs;
- use of appropriate technology for exploitation helps in conservation;
- adequate safety measures, health and welfare amenities are to be provided to the employees;
- with mutual understanding the employees can meet the need of the organization at a cheaper rate;
- it is beneficial to float the idea of equity participation in running a project;
- effective listening to the propositions put forward by the employee will play a major role in negotiations;
- they must allow the involvement of the employees from start-up to closure of a mine.

## Employee Action

It should be remembered by all the concerned parties that no gain can be achieved by one-sided involvement to establish the 'win/win' relationship. Therefore, employees, in their turn, must remember that

- a profit of the enterprise is their own profit and they must not leave any stone unturned to do their best;
- it is a rapturous perception on the part of an employee to obtain some real say in the work, supervision and monitoring process;
- while working in a team, ensure that your group has a recognition as an employee component in the organization;
- opportunities and skills achieved should be shared and transferred among the employees;
- business goal is distinct and deterministic and is different from political goal.

## Role of State

The State, being the first owner of the property and being the receiver of the 'royalty' and 'dead rent' should enter into an agreement with the entrepreneur, before the commencement of the mineral-property development, on the following basic principles:

— Easeness in processing and timely granting of the mining lease.
— Assurance from the developer/entrepreneur through EIA and EMP, that the development will not pose a threat of irreparable environmental damage.
— Keeping provision of services as basic facilities such as approach road, power, water, community market etc. where a large number of small-mines are expected to come up, without over-legislating
— basic educational facilities (a minimum of high school level) should be made available nearby, where the mineral-properties are located.
— Opportunities should be given to local people, which is very important—after all the local people will be on the land for 'time immemorial'.
— There should be provision for direct employment-oriented training and widest possible opportunity of employment for such trained local people.
— Opportunity should be created for the development of small-scale ancillary industries surrounding the project.
— Emphasis should be given on the development of rurally located mineral resources. This in turn, means the deveopment of the rural area, which directly contributes to the development of rural economy, a responsibility of the state.

## SLATE QUARRY—A CASE STUDY

In our country, many small-scale mines are operating in almost all the states. It is apprehended that some of them may not even have the concessionary rights; they may be evading the state inspection and continuing excavation flouting the basic principles of mining practices. In such small-mines, the workers work in inhuman conditions. With high accident risks, having no or little provisions for health services and welfare amenities, with meagre wages and under the conditions as good as bonded labour. A typical case was once published in one of the famous daily media. (Mojumdar, 1994) The excerpts form that in presented below.

On the lower reaches of Dhauladhar from Bhagsunag to Yole, about 8 kilometres from Dharamshala, are located hundreds of slate mines dotting the mountain slopes. The slate is culled from the rock surface. It is cut and fashioned into required sizes within the mines premises. The inhabitants of the closeby village Khaniara (the name derived from Khan — mines) and people from adjoining areas, earn their livelihood from these mines where they work as labourers. The slate mines in this area provide quick return on capital invested and form a lucrative business. The mines are leased out on an annual basis. The mines yield an annual turnover of Rs. 2–3 crores, providing direct employment to about 8000 people.

Traditional and unscientific quarrying methods are practiced, which has increased the hill slope to 89 degrees from 40 degrees original, thereby compounding the dangers. There exists little or no policy for employment and welfare for the workers. Apperently the workers have no prospects other than employment. The labourers work in a semi-bondage condition which is determined in the form of advance-loans taken from the contractors. Other employment opportunities are few in this area. Most of the mines are inaccessible by road. Mules are used for transportation of slate to the nearest market. Local populace suffer from chronic dust-related diseases due to lack of technical, safety and health measures. Tuberculosis, collection of collagen in the lungs reducing the lung capacity, Rheumatoid fever are common ailments. In most cases no trained mining engineer or blaster are employed in blasting process. In absence of blasting shelters and proper accessories for blasting, injury to the workers by flying rocks are very common. The heavily injured are laid-off by the contractors and have to spend the rest of their lives as dependants of their families.

To investigate and improve upon technical, safety, health and management conditions, governments due attention is drawn from time to time (Report, 1994). It is apprehended that such conditions may not be uncommon in other small-scale mining areas.

## CONCLUSIONS

Therefore, it is the time to solve such problems existing in small-scale mining activities in our country. Since it is an established fact that small-scale mining can provide a rewarding carrier to entrepreneurs and the employees which include operational, technical, scientific, financial, environmental and management opportunities and can be a source of smooth inflow of revenue to the exchequer, with 'win/win' relationship it is possible to achieve high standard of performance at a fair cost without flouting or evading the good practices of mining. Attention in addition, should be given to the health and welfare amenities of the workers, and the entrepreneur will soon realize the immense advantages derived therefrom. We trust that the deliberations will have desired effect. Through these ways the entrepreneurs and the employees thus, receive an opportunity to become successful in the venture in reaching their overall objectives of 'safe, economic and efficient mining' in small-scale endeavours.

### REFERENCES

1. Indian Minerals Yearbook 1991. Indian Bureau of Mines, Nagpur.
2. Small-scale Mining in India—Its Status and Perspective, 1984. prepared by MGMI, Calcutta.
3. Mukhopadhyay, S.K. and Ramlu, M.A. 1985. "Problems of Mining and Utilization of Small and Low Grade Mineral Deposits"; Proc. of Nat. Sem. on "Small Mineral Deposits—Their Development and Industrial Possibilities"; MGMI & FIMI, Nagpur, Nov., 4; pp. 70–78.

4. McAlpine, S. 1992. "Golden Bear Mine—Winning with First Nations"; CIM Bulletin, Vol. 85 No. 965, pp. 45–47.
5. Mojumdar, A. 1994. "Where Miners & Mountains Get A Raw Deal"; The Statesman, Calcutta, Apr., 3.
6. Report 1994. "More Powers for States in Mining Sector"; The Statesman, Calcutta, Mar., 17.

# MODERN EDUCATION AND TRAINING FOR SMALL-MEDIUM SCALE MINING MINERAL POLICY NEEDS

*M.B. Katz*
Director, Key Centre for Mines, UNSW Sydney, NSW, 2052, Australia

**INTRODUCTION**

The mineral development of emerging economic countries that have an important small-medium scale mining sector such as Vietnam, Laos, China, Mongolia, North Korea and countries like India as well as similar countries in Africa and Latin America have, in varying degrees, promising geological prospectivity and in the case of China and India an important mining industry. In regard to mining in GDP, of which the small-medium scale sector is important, Mongolia rates at about 40%; while China and India are about 50%; other countries even less so. In all cases the trend is to econmically expand this sector in the future. Although the geological potential is favourable for mineral exploration and development, other equally, if not more important, factors are taken into consideration by mineral investors concerned with investment risk. These include security of tenure, ability to repatriate profits, mineral policy, management control, mineral ownership and etc. Most of these countries are working on mineral legislation to take these factors into account, often without much success (e.g. Vietnam), so individual mineral project contract negotiation is common practice. Special problems exist in relation to artisanal small scale mining policy (Burke, 1996).

Thus many of these countries are exploring relevant mineral law models with their key personnel on various study and familiarization training tours to countries like Australia, which has the range of small-medium, up to very large mineral project development. Many of these training projects are funded by international funding agencies and are arranged with the aim of learning about various mining law models toward formulating appropriate national mining legislation. Wide ranging questions on state regulation, taxation, investment and environmental law are commonly raised by these technocrats and bureaucrats involved, that have little experience in these areas and also have needs to update and refresh their technological and management skills, especially business know-how.

## MINERAL POLICY FACTORS

Otto (1993) has described the various (Otto-Bakker) factors necessary for mining investment decisions that should be part of any small-medium scale mineral policy act. These include regulatory, fiscal, monetary and environmental criteria:

### Regulatory Crieteria

- workable mineral legislation
- stability of exploration/mining terms
- mineral ownership
- surface/land ownership or access to land
- security of tenure
- quality of mineral titles system
- right to tranfer ownership
- size of exploration blocks/duration of exploration rights
- availability of a mineral agreement in place of the mining code
- dispute resolution
- excessive bureaucratic intervention
- procedural efficiency and clarity.

### Fiscal Critteria

- tax method and level of tax levies
- ability to predetermine tax liability
- availability of tax holiday
- availability of accelerated depreciation
- availability of investment tax credits
- availability of reinvestment tax credits
- deduction of exploration costs
- export-import credits
- stability of fiscal regime
- tax treaty with home country.

### Monetary Criteria

- realistic foreign exchange regulations
- external accounts permitted
- ability to repatriate profits
- ability to raise external financing
- availability of local venture capital.

## Environmental Criteria

- legal requirements for environmental protection
- ability to predetermine environment-related obligation
- anti mining groups
- relative sensitivity of the environment.

## TRAINING PROGRAMS

The Key Centre for Mines International has been involved in various mineral policy familiarization and training programs since 1994 for Pakistan, China, Vietnam, Mongolia and Laos, which all have important small-medium scale mining sectors. The cooperation of the NSW Department of Mineral Resources and Mine Inspectorate is an essential component of the training. The training is essentially a familiarization exercise that examines the theory and practice of the mining act using the NSW regulations and experience as a guide. The NSW act is relevant to small-medium scale mining as there exists in the State special small scale mining regulations for gem and opal mining.

The trainees are first introduced to the mining regulations and operations by visiting the NSW Department of Mineral Resources to be briefed on the State Mining Act and Laws, including the Coal and Mine Inspectorate Branch. Arrangements are made at this time to organize mine inspection tours with regional State mine inspectors so that the trainees can observe these regulations put to practice.

Visits to mining companies reveal the company's approach to obtaining exploration leases and mining rights as well as environmental impact and other important issues like land rights and ownership. Visits to mines and mills also are important to observe company compliance. Academics and law firms with experience in mining law are also approached for input at various stages in the program. Relevant Key Centre for Mines short courses and special lectures are also included. The trainees are exposed to an appropriate mix of government and industry mining groups and organizations as well as the legal profession and the academia.

## KEY CENTRE FOR MINES TRAINING PROGRAM

An example of a 3-month program designed for a 1996 UNDP project for Lao PDR undertaken by the KCM is outlined below:

*1st Month*

1st week — Orientation and Introduction — KCM and UNSW
2nd week — NSW DMR and Mine Safety Laboratory
3rd week — KCM Short Course — Mineral Exploration Management
4th week — Small Scale Mining Field Study Tour

*2nd Month*

5th week — KCM Short Course — Computing and Statistics
6th week — Special Course — Mining Regulations
7th week — Small-Medium Scale Mine Inspection Tour
8th week — KCM Short Course — Economics, Commerce and Evaluation

*3rd Month*

9th week — Special Course — Environment Regulations
10th week — Mining Company Presentations and Visits
11th week — KCM Short Course — Project Valuation
12th week — Review, Report and Conclusions

## CONCLUSIONS

The worldwide competition for small-medium scale mineral development investment has resulted in the realization by many developing country policy makers that necessary favourable regulatory and fiscal systems must be established. Australia is in a good position to assist in this respect as the States and the Territory have enacted modern mining laws and regulations over the small-medium-large scales of present mineral development that could serve as appropriate models. The role of government and the mining industry and their relationships can be conveniently demonstrated. The developing country policy makers would benefit by familiarization and study training tours that would expose them to mining policy theory, development and application as practiced in State like NSW that are relevant to small-medium scale mining.

### REFERENCES

Burke, G. 1996. Policies for Samll Scale Mining: The Need for Integration, in Mining and Mineral Resource Policy Issues in Asia-Pacific — Prospects for the 21st Century D. Denoon, C. Ballard, G. Banks & P. Hancock (Ed.), ANU, Canberra, 103–106.

Otto, J.M. 1994. International Competition for Mineral Investment: Implication for the Asia-Pacific Region, in Asia Pacific Resource Development, P. Crowley (Ed), Proc. MEF PECC, Beijing, 11-24.

# MINING INFORMATION SYSTEM (MINIS) AND GLOBAL NETWORK

*Partha Pratim Chaudhuri
National Institute of Small Mines (NISM), Calcutta, India

## INTRODUCTION

International Development Research Centre (IDRC), Canada had commissioned Small Mining International (SMI), Canada and National Institute of Small Mines (NISM), India to develop an International (Small Scale) Mining Information System (MINIS). The duration of the project was for 5 years. The system, developed and tested with data on Indian context, has been so designed that it can be used to handle international data as well.

## NEED FOR MINING INFORMATION SYSTEM

Good documentation and reliable information are required to develop and strengthen the small mining sector which often can play a very useful role in disadvantaged communities by providing employment and higher standards of living. A well-designed information system in respect of a country will facilitate the exchange of information and experiences between countries and will assist governments, technical institutions and individual small miners to better understand, support, promote and operate small mines.

## OBJECTIVES OF MINING INFORMATION SYSTEM (MINIS)

The general objective of MINIS is to initiate the establishment of an international small-scale mining information system that would serve to facilitate the exchange of information and experience between countries, and include the development of a national information service in India.

### Specific Objectives

The specific objectives of the project are:
- to build a database management system on critical aspects of small-scale mining, with the capability of accommodating bibliographic, sta-

tistical, text and graphical/spatial data, as may be required;
- to provide a facility for assisting government departments, technical institutions and individual small miners alike to better understand, support, promote and/or operate small mines;
- to deliver basic information, analysis and replies to specific queries to potential users in a rapid and timely fashion and on a regular basis;
- to survey basic information needs of the small mines community in India and identify available information resources;
- to identify researchable issues on social and technical aspects of small-scale mining;
- to strengthen the capacities of SMI and NISM.

**FORMAT FOR MINE INFORMATION**

For the purpose of development of a structured format for data collection on Mining and other related Mine Information in respect of individual mines, initially a "need survey" was conducted through direct interaction with various small/medium mines. The task was undertaken primarily by the Research Assistant who had travelled to about 30 mines within the target area where he had one to one interaction with mine owners, mine managers and other persons associated with these mines.

Parallely, NISM had collected the sample copies of various returns which the mine owners are required to submit to statutory bodies like Indian Bureau of Mines, Directorate General of Mines Safety, State Mining Directorate.

Based on the outcome of the "need survey" and consulting the above return formats, a Data Collection format for mining has been developed. The format has been divided into the following 15 chapters:

1. General
2. Location of the Mine & Infrastructure
3. General Geology and Ore Mineralogy
4. Mine Production
5. Explosives
6. Employment—Manager/Technical & Mining Staff and Workmen
7. Equipment Used
8. Energy Consumption
9. Land Use & Environment
10. Sociological Aspects
11. Safety Aspects
12. Marketing
13. Financial Investment & Expenditure
14. Processing
15. Miscellaneous Problems, Queries & Assistance Needed

It may be noted that this format before finalisation had been discussed at length at the Project Executive Council. The draft format was also distributed to leading mining companies and reputed geo-mining professional within the country. Their comments on the format were obtained and changes were incorporated wherever necessary.

## TARGET AREA

Considering that the NISM Head Office is located at Calcutta, it was decided to concentrate on the mining area located in the tri-junction in the states of Bihar, Orissa and West Bengal. This has reduced logistics problems on travel, communication etc. It may be noted that this area has a number of clusters of mines and attempts were made to collect information from over 600 mines.

## MINING INFORMATION SYSTEM (MINIS)

After a long discussion at Montreal in October, 1992, between NISM and SMI personnel and subsequent visit of SMI Information Specialist to Calcutta and NISM Project Leader and Research Assistant to Montreal during November, 1994 and the discussion held then the decision was taken that the Information System (MINIS) will be developed at NISM end which may be shared and used through network.

Keeping this in mind the software MINIS started developing at NISM end and Duckback Information System of Calcutta was appointed as Consultant (Computer Information).

### Hardware Requirement

One 486-AT with at least 8 MB RAM and VGA Colour/Black & White SVGA Monitor
    HDD - At least 100Mb
    FDD - One 3.5" & preferably extra one 5.25"

### Minimum Software Requirement

1. Windows for Work Group (V-3-11)
2. FOXPRO 2.5 A for windows
3. MINIS Software
4. CDS/ISIS & WP 5.1 under DOS

## Application Software (MINIS)

The application software MINIS has been developed under FOXPRO 2.5 A under Windows. It is a multi-user environment software; it can be used under network.

### Databases Handled

The databases handled directly by MINIS are:
- Mining
- Equipment
- Sociology
- Corporate
- Consultant
- Financial
- IBM(Geo-mining data)
- Commodity

And the databases which are invoked through MINIS are:

Bibliography—this has been developed in CDS/ISIS
Legal—this has been kept in WP 5.1.

### 6.5 Modules

The entire software can be divided into 3 modules, namely:
(a) Data Entry Module—this deals with entry of data to all the databases discussed earlier.
(b) Reports & Queries—this module deals with all complex queries and reports. There are more than 250 standard RdQ in the System (MINIS). Other than this, provision for adhoc queries and reports are also kept.
(c) Utility:
   (1) Maintain some important master information which Common Users can utilise but cannot corrupt those information
   (2) Erase any information from Hard Disk and resort the database
   (3) Take backup and restoring of databases for future need
   (4) Print master information to find out any wrong information stored in the database

### User of MINIS

There are two types of user who can access this software

1. Common User
2. Super User

Common User (User Type—any Character) can handle only Data Entry Module

Super User (Type—S) can handle Data Entry Module as well as Reports & Queries and Utility

**Features of MINIS**

The features of MINIS are as follows:
- Maintenance of the databases for the different modules and provisions for Addition, Modification and Deletion of records from the database. A particular record can be searched from the database at any point of time.
- Generating reports from the database. Querying on the database to generate graphical outputs.
- Wrong codes (for some specific fields) at the time of entry are validated through HELP window.
- Calculator and unit conversion facility are also available.
- One can work on other environment (package or software) while working in MINIS by pressing CTRL + ESC key to invoke 'Take List' Dialog Box and then switching to Program Manager 'Window'.

**DATABASES**

The Mining Information System MINIS is a very detailed system and contains information under the following databases:

**Mining**

This database handles detail information on the following aspects of the mine:
- (i) General information of mine
- (ii) Information on infrastructure of the mine
- (iii) Information on general geology
- (iv) Information on mine production
- (v) Information on explosives used in the mine
- (vi) Information on employment in mine
- (vii) Information on equipment used in mine
- (viii) Information on energy consumed for mining operation
- (ix) Information on safety aspects of the mine
- (x) Information on sociological aspects of the mine
- (xi) Information on landuse & environment of the mine
- (xii) Information on marketing of the products
- (xiii) Information on financial investment and expenditure
- (xiv) Information on processing of the product

(xv) Information on problems/querries and assistance needed by the small mine owners.

## Corporate

This database handles the following information of the corporate sectors dealing with minerals and mines:

 (i) Name and address of the company
 (ii) Location of mines (if any)
 (iii) Location of processing unit
 (iv) Activities of the company such as:
  - mining
  - manufacturing & processing
  - export
  - trading
  - minerals handled
 (v) Information on associate companies

## Commodity

This database deals with information regarding marketing & processing of a particular mineral. The information heads are as follows:

 (i) Specification & Grade
  - National
  - International
 (ii) Occurrence
  - National
  - International
 (iii) Production Trends
  - National
  - International
 (iv) Mineral Processing Techniques
 (v) Description of Processing Methods
 (vi) Experience on Beneficiation
 (vii) Price trends
  - National
  - International
 (ix) Market Mechanism
 (x) Export Market
 (xi) Major User Industries

## Consultant

This database handles information of individual, groups or companies who are in the field of mining & mining related consultancy and contains information under following heads.

  (i) Name and Address of the Consultant (individual or organisation)
 (ii) Area of Specialisation
(iii) Practical Specialisation
(iv) Professional Qualification
 (v) Membership of Professional Association
(vi) Professional Assignments.

## Equipment

This database mainly deals with equipment manufacturers and details of their products. The information are kept under following heads:

 (i) Company details
(ii) Model details including scanned images and specification of the models.

## Financial

This database deals with the financial institutions involved in financing in mining & mining related fields. Information stored are:

  (i) Financial institution details
 (ii) Available schemes of financial assistance for mining, mineral processing and allied industries
(iii) Loan application procedure
(iv) Location of regional and branch offices.

## Sociology

This database deals mainly with socio-economic and socio-cultural aspects in the mining area. Information gathered under this head are mainly on

  (i) Household basis
 (ii) Village level
(iii) Tribal women in particular

### Legal

Services of reputed professionals in the field were utilised for the development of a write-up relating to various statutory provisions on mining. This covers items like obtaining mining lease, their renewals, working provisions and other related matters.

### Bibliography

Bibliography database has been developed utilising CDS/ISIS software.

## FUTURE ACTIVITY WITH THE SYSTEM—GLOBAL NETWORK

The system has been so developed that it can be used under network and shared by different countries simultaneously on line. The system can also handle data of different countries.

Our expectation is to install the system in different countries, which would be interlinked with each other as well as with NISM and SMI through internet. This connectivity will be an on-line one.

The nature of activity will be as follows:

(a) NISM and SMI would install the software MINIS in different African, Asian & South American countries and would train the personnel regarding handling and maintaining the system and the data bases.
(b) NISM & SMI would guide the institutions/countries, who would install the software, regarding connectivity between them and NISM & SMI.
(c) Countries/institutions including NISM, would feed the system with data of their own countries which could be retrieved through the system at any point of time as and when required. Thus each of the countries/institutions including NISM & SMI, would function as independent connected nodes of the system thus globalising the system.
(d) From any of these nodes any specific question from outside MINIS also can be fired and can get back the required information. Similar things can also be added to the system e.g. the recent policy changes on any mining aspect can be asked for from any country which can very easily be answered, though not through the system.
(e) The major task at each node will be to gather data for the nodal data bank, as well as to update them in a regular manner.
(f) NISM & SMI would maintain the system, incorporate the required changes as and when required and simultaneously provide them to the different nodes.

## FUTURE OBJECTIVES

(i) To globalise the Mining Information System for possible investors.
(ii) Maintain information databases on different aspects of mining in different countries.
(iii) Build up chain world wide for gathering and dissemination of information for interchange of ideas and experience.
(iv) Build up a network with SMI as global apex body and different countries/ institutions as continental apex bodies (for example India as an apex body for Asia or Zimbabwe as apex body for Africa, Brazil for South America etc.).

## EXPECTED OUTPUT

The expected outputs are:

(i) Information gathering and dissemination both nationally and internationally.
(ii) Transfer of low cost appropriate technology in the field of mining equipment, exploration techniques, mineral processing etc.
(iii) Provide the small mine owners with the information regarding market, finance, low cost beneficiation etc both nationally as well as internationally.
(iv) In different countries, for example India, small mines are presently under the organised sector and the existing legislations and laws can form a guideline for the countries where they are going to bring artisanal and very small mines under the organised sector. Information on this aspect can also be obtained in details through this system and the network.
(v) The system will provide information of different financial institutions which the different nodes can axcess and pass on to interested persons/organisations/small miners.
(vi) Foreign multinationals interested in ventures or joint ventures in the field of mining/mineral industry can also retrieve information in the fields of their interest. They can also identify consultants who may be able to serve as useful contact point in any country
(vii) Preliminary information on socio-economic aspects necessary for starting mining operations in any country can also be obtained and disseminated appropriately.

## MEASUREABLE OR APPRECIABLE EXPECTED IMPACT

(a) Increased national and international investment in the field of mining or mineral related industries.

(b) Increase in employment in mining and mineral related industries.
(c) Improved mining standard resulting in better labour earning and eco-friendly operation consistent with safety.

**BENEFICIARIES**

The direct beneficiaries of the information services, provided internationally by SMI/NISM, will include the national small mines information centres when established; government agencies working to support, rationalize and expand national small-scale mining sectors; isolated small mine operators who are unable to link up easily with potential investors, consultants, equipment suppliers or to find answers to technical, financial or legal questions on a timely basis; private entrepreneurs looking for new international opportunities.

The users of the information services provided by NISM and SMI are expected to be many and varied. Included are small miners, through the interaction with local and international field workers, researchers, professionals, academics, government planners, international/national aid agencies, railways, public and private sector companies, manufacturers, small-scale entrepreneurs, and students.

**ACKNOWLEDGEMENT**

The author, one of the team members of the project, would like to thank IDRC, Canada for fully funding the project. He is grateful to the co-workers, Mr. Sandip Sarkar, Information Specialist; Mr. Ujjal Bandopadhaya, Research Associate and Mr. S.L. Chakravorty, Project Leader of NISM for helping him and asking him to prepare this paper. He would also like to thank the Directors of State Governments of Orissa, Bihar and West Bengal and also other Govt. officials for giving him unstinted support in collection of information and data for the project. He is also grateful to many of the Senior Geologist and Mining Engineers, namely Mr. Binoy Neogy, Mr. D. Lahiri, Mr. H.B. Ghosh for thier valuable advice in preparing and designing the different data collection formats.